Sick Building Syndrome and Related Illness

Prevention and Remediation of Mold Contamination

Sick Building Syndrome and Related Illness

Prevention and Remediation of Mold Contamination

Edited by
Walter E. Goldstein, Ph.D, PE

CRC Press
Taylor & Francis Group
Boca Raton London New York

CRC Press is an imprint of the
Taylor & Francis Group, an **informa** business

CRC Press
Taylor & Francis Group
6000 Broken Sound Parkway NW, Suite 300
Boca Raton, FL 33487-2742

Printed in the United States of America on acid-free paper
10 9 8 7 6 5 4 3 2 1

International Standard Book Number: 978-1-4398-0144-4 (Hardback)

Library of Congress Cataloging-in-Publication Data

Sick building syndrome and related illness : prevention and remediation of mold contamination / editor, Walter E. Goldstein.
p. cm.
Includes bibliographical references and index.
ISBN 978-1-4398-0144-4 (hardcover : alk. paper)
1. Sick building syndrome. 2. Molds (Fungi)--Health aspects. I. Goldstein, Walter E. II. Title.

RA577.5.S517 2011
613'.5--dc22 2010023846

Visit the Taylor & Francis Web site at
http://www.taylorandfrancis.com

and the CRC Press Web site at
http://www.crcpress.com

Dedication

This book is dedicated to providing information and ideas that lead to progress to help those who suffer from effects of sick building syndrome. The authors hope that this book will have a role in alleviating suffering to improve health and well-being.

To my wife, Paula, for her steadfast love and support of my professional efforts, even now as I am in the twilight of my still-active professional career.

To my wonderful children, Susan and Marc, and Marc's loving wife, Sunita, and our wonderful grandchildren, Alex, Noah, Reena, and Daniel, for the love, pleasure, and happiness they have provided to us.

Contents

Preface

Sick building syndrome and attendant health problems are a widespread disaster-in-the-making that threatens our national economy. A sick building is a structure that has become contaminated with any number of harmful agents. In this book, we focus on the serious contaminants causing fungal infestation, commonly referred to as mold. Mold may be allergenic in itself, its components may be allergenic, and the contaminating organism may also exude allergens and deadly toxins.

The problem occurs worldwide. This is a tragic matter since the most seriously afflicted are often children due to the asthma that may result. Mold plagues the richest and poorest of homes and people alike. The cost to health is significant. A water leak can occur behind a wall in a dwelling, perhaps unknown to its occupants, until the leak appears or the sinister color of mold is evident. The occupants thereby become vulnerable economically and their health may be compromised. The cost to remediate or correct problems has the potential to ruin people financially unless they have an outstanding insurance company willing to cover such problems.

The problem of detection, observation, measurement, and remediation is presently treated in a reactive manner. A problem occurs then trained personnel are called in to provide their expertise. Just suppose that instead of being reactive, one could be proactive and predict problems and cause measures to be taken to counter problems before they occur (or sometimes after they occur but before they become injurious to excess). If this book is successful, it will provide a means to help in present-day, immediate, short-term solutions for existing dwellings affected with mold. It will also provide solutions for future construction so that the problem may diminish with time.

The book is organized to present the overall challenges involved in the subject, the history of building mold contamination, and the input of experts (scientific, architectural, engineering, health care, and environmental). Areas not covered as expertise (legal and regulatory areas) are presented in the form of practical suggestions from the experience of the authors in order to provide suggestions to those affected by mold or those working in the area.

The book is designed to be a reference with practical information for many groups, including consumers, remediation specialists, construction experts, scientists, engineers, attorneys, physicians, pharmacologists, psychiatrists, insurance specialists, government agencies, environmental groups, those providing short- and long-term education in the field, and the housing industry. Numerous references are provided to aid further research and investigation.

The book provides suggested solutions in each topic area plus a summary chapter on recommendations to address the interests of many parties. The recommendations encompass consumer issues and health care subjects. The health care subjects include aspects of medical and psychiatric matters and details from experts in epidemiological studies. Construction industry issues, insurance industry needs, regulatory matters, practical sampling of contaminated areas, testing, remediation, "put back" of building components postmold, and structural component removal are addressed. Special sections on advice to consumers and suggestions for new building construction are included.

The book addresses practical challenges, and the authors attempt to impart imaginative and creative input to bring fresh thinking to a difficult area. The text presents information on needs and concerns, plus suggested solutions appropriate to industrial, consumer, academic, and governmental sectors in the United States and worldwide. The book addresses many technical subjects in science, engineering, and medicine since these are critical to identifying solutions to sick and damp building syndrome challenges.

Better materials science and the ability to know when mold will occur and how to prevent and remediate it are suggested to be critical and key remedies to mold infestation. Sound science and engineering can be incorporated as a package as part of a home or commercial buyer's purchase. For example, the model for mold growth presented in this book (Chapter 5) can be adapted commercially to depict how mold growth can occur and how to prevent such growth.

Chapters on mold biology and metrics can inform parties how to approach measuring infestation and understanding it. Chapters on mold and other contaminant particles, remediation, and repair can provide insight to parties beset with mold problems to help them know what to do in the event of a problem. The chapter on epidemiology conveys an understanding of the problem and its magnitude and presents aspects of health challenges.

Key features of the book include:

1. Equations of mold growth and product formation useful in building design, mold prevention, and directing research to new solutions
2. Mold genetics, biology, and products
3. Mold epidemiology
4. Analytical developments and sampling techniques to measure mold infestation and products to help investigators and mold remediators
5. Photographs of mold infestation compared to other contaminant infestation
6. Guideposts in remediation and put-back
7. Case histories in mold remediation, including photographs
8. Practical consumer and construction industry guidance
9. Information to help attorneys and judges in legal cases

The parties that will benefit from buying and reading this book include the following:

1. Insurance industry (reference on claims)
2. U.S. Department of Housing and Urban Development (HUD; reference on new and existing housing with mold infestation)
3. Centers for Disease Control and Prevention (CDC; disease and health considerations due to mold)
4. Housing and construction industry nationwide (optimizing construction, steps for developers dealing with this issue and liability)

5. Academic scientists and engineers (looking for grants in this area)
6. Consumers with mold and health problems they attribute to mold
7. Industrial scientists and engineers seeking uses for their products
8. Physicians and psychiatrists who have to treat patients and wish background information on health care, and pharmacologists developing therapeutic agents to treat and prevent mold-related diseases or symptoms
9. Attorneys and judges who must deal with legal cases and lawsuits
10. Educators who wish to teach the subject and communicate findings to the community

Acknowledgments

To Peter Alexander, PhD, who first approached me to develop what became a successful symposium on this topic held at the American Academy of Forensic Sciences 2008 meeting.

To Richard C. Lee, PhD, planned contributor, who suddenly passed away in 2009, former vice provost and head of Educational Outreach at University of Nevada, Las Vegas (UNLV), for his encouragement and support of this effort, general friendship, and as a considerate and able supervisor in my final year at UNLV.

To Taylor & Francis, CRC Press publishers for agreeing to issue this book, and in particular, Jennifer Ahringer, project coordinator, and Joseph Clements, acquisitions editor, for their support, advice, and actions to help ensure that this publication occurred.

To my colleagues and coauthors, all experts in the subject matter, who followed through and provided their chapters in a shared belief in the importance of this topic and its potential contribution to meeting national and world challenges in this area.

Editor

Walter E. Goldstein, PhD, PE is president of Goldstein Consulting Company (www.goldconsul.com), offering services in chemical engineering and biotechnology. He is called upon to improve processes and products in health care, consumer products, food, pharmaceuticals, chemicals, and biofuels. He provides expertise to model and analyze product and process components, changes, and defects; construction practice challenges; and alterations to materials, valuable artifacts, and memorabilia. Many of his projects involve biotechnology processes that include mammalian cell, bacterial, and mold propagation. He has extensive background in fermentation technology and development of products and processes in this field. He is also president and cofounder of TransCyte Inc., a research and development firm with an issued patent and objective to produce universal blood from stem cells in a bioreactor process thereby avoiding use of donor blood for transfusion.

Goldstein was vice president for biotechnology research for Miles Inc., a former division of Bayer Inc., from 1982 to 1987. He was also vice president and director of research for ESCAgenetics Corporation, a plant sciences biotechnology company, from 1988 to 1994. He founded Goldstein Consulting Company in 1994 and has been engaged in several entrepreneurial enterprises since that time.

He founded and developed a forensic sciences DNA profiling laboratory at the University of Nevada, Las Vegas from 2003 to 2008. It was during this period that the concept for this book was initially developed. He organized a symposium on sick building syndrome in 2008 at the request of principals of the American Academy of Forensic Sciences.

Goldstein holds a BS degree in chemical engineering from the Illinois Institute of Technology (1961), and MS and PhD degrees in chemical engineering from the University of Notre Dame (1973). He also holds an MBA from Michigan State University (1968) and is a Registered Professional Engineer (PE). He is a member of the American Academy of Forensic Sciences, the American Institute of Chemical Engineers, and Sigma Xi.

Goldstein resides in Las Vegas, Nevada, with his wife Paula. They have two children, Susan and Marc, and four grandchildren, Alex, Noah, Reena, and Daniel.

Contributors

Gary R. Brown, PE, QEP, CMC
RT Environmental Services, Inc.
King of Prussia, PA

Brad Caddick, ASP
Environ International Corporation
Chicago, IL

Michael J. Cornwell, AIA
Environ International Corporation
Chicago, IL

Jean Cox-Ganser, PhD
Centers for Disease Control and Prevention, National Institute for Occupational Safety and Health
Washington, DC

Gary N. Crawford, CIH
Environ International Corporation
Chicago, IL

Barb Epstien, MPH, CIH
Epstien Environmental Resources, LLC
Marietta, GA

Walter E. Goldstein, PhD, PE
Goldstein Consulting Company
Las Vegas, NV

Elliott Horner, PhD
Air Quality Sciences, Inc.
Atlanta, GA

Richard Kanwal, MD, MPH
Centers for Disease Control and Prevention,
National Institute for Occupational Safety and Health
Washington, DC

James Millette, PhD
MVA Scientific Consultants, Inc.
Duluth, GA 30096

Philip R. Morey, PhD, CIH
Environ International Corporation
Gettysburg, PA

Ju-Hyeong Park, ScD
Centers for Disease Control and Prevention,
National Institute for Occupational Safety and Health
Washington, DC

Willem A. Schreuder, PhD
Principia Mathematica
Lakewood, CO

Edward Sobek, PhD
Assured Biotechnologies Corporation
Oak Ridge, TN

Tara Toren-Rudisill, AIA
Environ International Corporation
Chicago, IL

Raoul A. Webb, PE
Environ International Corporation
Tampa, FL

1 Introduction

Walter E. Goldstein, PhD, PE

CONTENTS

1.1 PREAMBLE

Mold (or fungal) contamination has been with us for quite some time. It became a more significant problem after WWII when construction practices changed to supply and use of less expensive materials. Circumstances conducive to mold growth in buildings and associated problems subsequently arose.

Sick building syndrome and attendant health problems are a widespread disaster-in-the-making that threatens our national economy. A sick building is a structure that has become contaminated with any number of harmful agents (Institute of Medicine, 2004, 250). They are many aspects to "a sick building" (U.S. Environmental Protection Agency, 2008). In this book, we are focusing on the serious contaminants causing fungal infestation, commonly referred to as mold. Mold may be allergenic in itself, its components may be allergenic, and the contaminating organism may also exude allergens and deadly toxins (Institute of Medicine, 2004, 58).

The problem occurs worldwide (Gutarowska and Piotrowska, 2007). This is a tragic matter since the most seriously afflicted are often children due to the asthma that may result (President's Task Force on Health Risks and Safety Risks to Children, 2000). Some remedies proposed by the Presidential Task Force on Health Risks and Safety Risks to Children (2000) may well be implemented by studies as reported in the literature (Kercsmar et al., 2006).

Mold plagues the richest and poorest of homes and people alike. Mold has been reported at least since the time of Moses (Sedlbauer, 2002a, 5). Often, present-day remediation practices still incorporate what was done in ancient times. The cost to health is significant. For example, the proportion of the cost of mold related to asthma illness is estimated by one party to be $3.5 billion in the United States (Mudarri and Fisk, 2007). The cost to remediate or correct problems has the potential to ruin people financially unless they have an outstanding insurance company willing to cover such problems.

A water leak can occur behind a wall in a dwelling, perhaps unknown to its occupants until the leak appears or the sinister color of mold is evident. The occupants thereby become vulnerable economically and their health may be compromised. One calls a contractor to correct and remediate the problem. The attendant costs that become known are often staggering to an individual. If the problem due to water leakage or mold remediation is covered by insurance, coverage for mold is perhaps limited.

Even if the insurance company agrees to cover the cost, it is vulnerable and may become insolvent or before that, if well managed, simply refuse to provide insurance if too many cases occur. A consequence of too many decisions of noncoverage is to thereby destroy the personal well-being of a significant fraction of the U.S. population. Even if the financial well-being is manageable, the danger to the health of those affected and sensitive to the mold infestation is severe from a humane standpoint, adding serious burden to our already overburdened health care system (Mudarri and Fisk, 2007).

The problems of detection, observation, measurement, and remediation are presently treated in a reactive manner. A problem occurs and trained personnel are then called in to provide their expertise. Just suppose that instead of being reactive, one could be proactive and predict problems and cause measures to be taken to counter problems before they occur (or sometimes after they occur but before they become injurious to excess).

If this book is successful, it will provide a means to help in present-day, immediate, short-term solutions for existing dwellings affected with mold. It will also provide solutions for future new construction so that the problem may diminish with time. In Chapter 5, for example, a framework of mathematical modeling is presented, that, if developed and supported by a grant award or other means, will encompass the significant elements of mold research and field investigation, so that the model and its underpinning experimental data and associated analytical methods could be the basis for the derivation of solutions in a predictive and dynamically reactive manner—in effect, producing a valuable expert system. Certainly, such work will directly address a multitude of needs covered in government policy documents (Department of Housing and Urban Development, 2008, 3) and have practical beneficial implications at the local level.

A symposium, "Sick Building Syndrome," was held at the American Academy of Forensic Sciences meeting in Washington, DC on February 21 and 22, 2008 (Goldstein et al., 2008). The symposium was organized to (1) report on sick building syndrome challenges, (2) seek corrective actions, and (3) derive short- and long-term solutions to protect those affected. The symposium theme addressed basic causes of the problems and provided some suggested (speculative) solutions in this complex area. The success of the symposium provided the impetus for this book—to follow up (and expand) on recommended suggestions and potential solutions.

The book is organized to present the overall challenges involved in the subject, the history of building mold contamination, and the input of experts (scientific, architectural, engineering, health care, and environmental). Areas not covered as expertise (legal and regulatory areas) are presented in the form of practical suggestions from the experience of the authors in order to provide suggestions to those affected by mold or those working in the area. The book is designed to be a reference with practical information for many groups, including consumers, remediation specialists, construction experts, scientists, engineers, attorneys, physicians, pharmacologists, psychiatrists, insurance specialists, government agencies, environmental groups, those providing short- and long-term education in the field, and the housing industry. Numerous references are provided to aid further research and investigation.

The book provides suggested solutions in each topic area plus a summary chapter on recommendations to address the interests of many parties. The recommendations encompass consumer issues and health care subjects. The health care subjects include aspects of medical and psychiatric matters and details from experts in epidemiological studies. Construction industry issues, insurance industry needs, regulatory matters, practical sampling of contaminated areas, testing, remediation, "put back" of building components postmold, and structural component removal are addressed. Special sections on advice to consumers and suggestions for new building construction are included

The book addresses practical challenges, and the authors attempt to impart imaginative and creative input to bring fresh thinking to a difficult area. The text presents information on needs and concerns, plus suggested solutions appropriate to industrial, consumer, academic, and governmental sectors in the United States and worldwide. The book addresses many technical subjects in science, engineering, and medicine since these are critical to identifying solutions to sick and damp building syndrome challenges.

Better materials science and the ability to know when mold will occur and how to prevent and remediate it are suggested to be critical and the key remedies to mold infestation. Sound science and engineering can be incorporated as a package as part of a home or commercial buyer's purchase. For example, the model for mold growth presented in this book (Chapter 5) can be adapted commercially to depict how mold growth can occur and how to prevent such growth. Chapters on mold biology and metrics can inform parties how to approach measuring infestation and understanding it. Chapters on mold and other contaminant particles, remediation, and repair can provide insight to parties beset with mold problems to know what to do in the event of a problem. The chapter on epidemiology conveys an understanding of the problem and its magnitude and presents aspects of health challenges.

1.2 FINANCIAL IMPACT ON CONSUMERS, BUSINESS, AND THE NATION

This section of the introduction presents background information on examples of the cost of mold to an individual and nationally. It addresses true impact of mold and consequences. It is clear that people faced with mold infestation find their fortune and life in the hands of others. The national cost is the sum of the cost to many individuals such as homeowners. This cost can involve their dwelling or business, and can involve the cost to cure a health problem due to mold for themselves, family, or employees. It was mentioned already that the cost contribution to asthma illness due to mold is estimated to be $3.5 billion (Mudarri and Fisk, 2007).

The construction industry is affected since requirements to avoid mold or to respond to mold infestation charged to be their responsibility can add excessively to costs plus deter them from building or adding value to the economy if the risk involving mold is too high for them to incur. The insurance industry is affected since it has to decide whom (and for what) to insure and how to compensate in a consistent and rational manner. Also, the insurance industry must decide what premiums to charge to insure the industry is protected financially.

The propensity of mold to develop, the time factors, and the materials used that can resist or promote mold infestation affect an insurance actuary's work and his or her models to predict losses and to recommend premiums. The same subjects affect the builder and developer since this individual must decide which materials to use in construction and if construction is indeed feasible. Builders and developers must assess their risks and whether a project is financially viable, short and long term.

Knowledge of mold growth and the ability to predict the likelihood of mold to grow under different circumstances would be valued. The value is much greater (for example, to the construction and insurance industries) if the knowledge leads to understanding the most economical way to treat and prevent mold growth, and to remediate a structure that had become moldy and unsafe. Providing such information to these industries and directly to consumers thereby provides protection and benefits the economy. Presenting this knowledge is the main theme of this book.

The national cost due to mold is comprised of many factors. The cost to health when mold causes allergies, breathing problems, and psychological problems (leading to psychiatric therapy needs, for example) may be quite large, especially if one develops debilitating illnesses that have a lifelong impact.

For the average consumer, a response to discovery of a mold problem can be a matter of notifying competent and reliable technical specialists, while simultaneously tapping into good insurance. For the less fortunate, this problem can involve dealing with a situation where technical experts that inspire confidence and are reliable are not available, and where insurance is inadequate. For example, some mold policies may be limited to $5,000 when the cost of remediation could be many times that. Often an insurance company will not insure beyond a minimal amount, for example, $5,000, even if desired and requested. Often, the extent of the coverage is hidden from the average consumer and not explained clearly in policies. The government (federal, state, and local) may be of little use in this situation unless statutes are clear, and they are not in this case, nor perhaps are such statutes adequately in place. If the statutes are in place, what they offer may be limited and of little value. However, government at all levels is working to improve this area even if desirable remedies do not extend to local levels.

Those who are poor and in substandard housing, perhaps subject to less desirable and unethical landlords, can be victims. Those in standard and above-standard housing can still incur problems with mold due to shoddy construction and basic unawareness of a problem that may be preordained before a new house is occupied or may occur due to a sudden event such as a water leak.

The water leak may be accidental or due to a plumbing failure. The plumbing failure may be due to poor workmanship or choice by the contractor of plumbing materials inappropriate for an area, for example, due to corrosion or structural weakness causing failure. Such contractor decisions may be intentional to save money or simply an accident and may be an indication of incompetence. An example of poor material selection is selection of copper pipe for use in an area where the ground contains a high concentration of available ammonia. Ammonia will react with copper and destroy it.

The question then remains: What is a consumer to do? First, there are services one can retain to check a dwelling for water intrusion and mold before signing a contract to purchase. In areas with good laws and regulations, this is required. In some areas, due to oversight and the backward nature of local legislation, this may have to be specially requested. The consumer and commercial occupier must be proactive and check these things in advance. This can be done by the purchasing party, or someone can be hired such as a professional inspector or licensed professional engineer.

Tests for water leaks can be done nonintrusively through walls. Direct tests for mold presence can also be made at greater costs and are intrusive (wall penetration is required to obtain a sample). However, such tests may be inadequate or not equal to the task. For example, detecting an organism requires it being present in sufficient quantity to collect it, or to detect an organism byproduct, or collect the organism to isolate its DNA. Sometimes, the technology is not sufficient for this task if the organism only exists, for example, as a spore, which may be ubiquitous in any case and just waiting for water intrusion to cause it to actively propagate (germinate much like a seed changes to plant form, and then grow in order to do its damage). The occupancy of a dwelling by a buyer then becomes an investigative and science project. This involves added and necessary cost. Each person has to decide what is necessary and unnecessary. People have to take responsibility to make sure such matters are acceptable for them. This may involve relying on others and trusting others to follow through and provide the services needed. A mold insurance policy against devastating mold events taken early and applied to errors in inspection may be part of the answer if available. However, that involves cost and risk again.

If this area is unnecessarily overregulated, this will simply drive developers and contractors away and create a shortage of adequate new housing and skilled persons to correct housing deficiencies and needs. There was talk of a national insurance program that may have died in the U.S. Congress when the Melina Bill, originally introduced by Representative John Conyers of Michigan, died in committee (Conyers, 2003). A national insurance policy sounds like an answer. However, such a one-size-fits-all, cure-all without adequate local cooperation and legislation is simply not useful. Federal support in this area is desired, including funding for programs. The best legislation comes locally and is adapted to local needs. In other words, federal funding is desirable to get action started in this important area. However, excessive federal governance is not desirable or useful in regard to implementing spending of funds. The mechanism for proper spending should be in place locally.

Mold seems to habitually start with water intrusion (liquid or even gaseous, as humid air can cause some molds to propagate). Once material is wet and the air is ubiquitous with mold spores, the spores alight and begin their serious infestation. So clearly, the first line of defense, economically, is to discover water intrusion. Such intrusion may be obvious as if one spots a leak. However, such intrusion can be hidden behind a wall and not noticed until the color of mold penetrates the wall or people start to become ill, and someone traces this to mold (if the illness can be traced to mold by any standpoint, inspection or medically). Because water may be the culprit, getting a claim for mold may be dicey if parties key in on water first. This can of course have legal implications depending how statutes are written. In one case, a court refused to hear a claim in regard to mold since it had already reviewed and ruled on a claim for water damage. So in this case, will the insurance company consider the entire problem, including that of mold, water damage, and cover costs? Or, will they only cover strict water damage measures and leave the cost of mold remediation to the consumer. Since the insurance clauses may not be uniformly written or clear on this, the consumer should probably ask these questions up front and try to get a reply in writing in advance before purchasing a dwelling. However, getting the insurance company to reply in a manner that obligates it may not be possible. So when one is shopping for insurance, perhaps it is useful to determine which company may be most likely to honor such a complex claim, the premium it requires for this, and if it is less likely to drop you unexpectedly as a client following a misfortune with mold. Assessing this is a matter of personal judgment, supported by the sound advice of specialists and knowledgeable friends.

1.3 REGULATORY IMPACT

In a sense, there really isn't any regulatory impact since this area is largely unregulated. Bills introduced in Congress (e.g., Conyers, 2003) recently died in committee. Recommendations have been made and documents have been issued by the U.S. Environmental Protection Agency (EPA), U.S. Department of Housing and Urban Development (HUD; 2008), and the Occupational Safety and Health Administration (OSHA; 1970, 2008). However, all are recommendations, and nothing is binding. State OSHA agencies may strictly follow national OSHA guidelines. Technical guidance for the local level seems limited to suggesting training and licensing needs. This is window dressing since it is unclear that all such trainees understand how to apply existing science and engineering to the problem. Just because someone has a license or certificate does not mean he or she is any good and can help you. The consumers are basically on their own.

One cannot overregulate since passing greatly worded laws that cannot be implemented is harmful. Regulations that come at the national, state, and local level must be consistent with each other. Obviously, this may not occur for many reasons.

The EPA Web site has the following statement: "Standards or Threshold Limit Values (TLVs) for airborne concentrations of mold, or mold spores, have not been set. Currently, there are no EPA regulations or standards for airborne mold contaminants." Without such a standard, it is difficult to pass laws. The standards are not set probably because the science needed to set the standards is not sufficiently refined. Without any federal standard, it is expected that state and local standards may not be set as well. However, there are exceptions. California, a state noted for progressiveness in environmental areas, passed its toxic mold protection act in October 2001. Texas has apparently passed some regulatory statutes. Nevada may as well. However, local incidents in 2002 in Las Vegas, Nevada; New York City; and Texas, as examples, indicate that local regulations in regard to mold may be insufficient (Mummulo, 2009; Sherman, 2007; Smith, 2002; Whited, 2008).

Certainly, mold can proliferate in a dry climate like Las Vegas as it can in a moist climate such as that in Oregon. The spores in a dry climate may be more virulent and voracious. Regulations should be practical and not inhibitory to commerce. However, they must exist to provide basis protection for individuals. It would seem that statutes are needed locally, at the state level, and at the federal level. It makes most sense for the statutes to be developed locally and then be adapted at the state level and finally at the federal level to best serve citizenry, and not legislate before there are means to implement. The writings in this book can provide a basis for such laws since the phenomenon of mold growth is one based on science. All chapters in this book present essential science and engineering input that can be valuable in developing legislation.

One can purchase insurance. However, unless one is very astute, applying the insurance to protect oneself in the event of a mold calamity is very uncertain. It is a real situation of *let the consumer and buyer beware* (of health and financial risks). In such a complex claim area, one has to consider the cost of premiums to get desired protection, and if the particular insurance company is less likely to drop you unexpectedly as a client following a misfortune with mold.

1.4 MOLD-RELATED MEDICAL AND PSYCHIATRIC ILLNESS, PHARMACOLOGICAL SOLUTIONS, AND PATIENT ISSUES

Medical and epidemiology issues are addressed in Chapter 2. Other chapters present aspects of health issues involving mold. Often, one does not know if an issue is a medical issue, a psychiatric issue (a real effect or one imagined, perhaps manifested physically), nor is it certain which drugs to apply to treat mold or treat the consequences of mold on human health. There are many patient issues where people are seriously affected. There are also charlatans trying to gain financially and illegally. Sorting this out is very difficult.

A person exposed to sick building syndrome may be ill due to the presence of the mold or the perceived presence of the mold. The mold organism, cells or spores, and mold products (biochemicals) can all cause real health issues that perhaps are sustained. A physician treats symptoms and observes results. If a person is in a mold-infested environment, the symptoms will persist and may be deadly. Perhaps they can be diagnosed. Lingering issues may be psychiatric in nature still requiring medical treatment. Those affected may seek legal remedies (which spills this over into the legal area). Suits may be civil. If a malevolent act is involved, then the legal matter may involve a crime.

Given a mold-related illness is incurred, parties affected may manifest physiological symptoms, behavioral symptoms, or both. The symptoms may be real or psychosomatic. In the latter case, they are still real. The physician, possibly psychiatrist, must find ways to treat the patient, which may include counseling or pharmaceutical therapy or both. It would be desirable if the diagnostics used could pinpoint the microbial source of the problem including any mold secondary metabolite (mold product) cause. It would then perhaps be possible to better attempt therapy and be able to record results, good or bad, in regard to treating the patient. Very quick, accurate, and precise tests would be helpful here, particularly if readily accessible to the physician, psychiatrist, and the diagnostic laboratory in use by these professionals. Chapters 3 and 9 cover mold biology and practical tests based on mold genomics. Such tests may become a basis for improved clinical tests that would help in defining therapy.

The availability of such testing technology is desirable since the literature notes the many molds and the many secondary metabolites that can cause problems (and associated symptoms). Health issues are reviewed extensively in Chapter 2. Practical information on this subject is also provided in Chapters 4, 6, 7, and 8. Unfortunately, as yet there are no sufficient information bases to relate the affliction symptom to the mold or mold product, and then to the attempted cure.

Science can play a role by providing better diagnostics to detect the mold and its products as noted in this book. Science and medical research can also provide better therapeutic agents to treat symptoms. A mold infestation that becomes systemic can be treated with an antifungal therapeutic agent, taking care to ensure that severe byproduct reactions do not result.

A sick building may contain a complex mixture of infectious bacteria, fungi, endotoxins (from bacterial cell walls), mycotoxins (from fungal cell walls and also exudates), and particles causing irritation to tissue (for example, lungs, airway passages, skin, internal organs if ingested). The attending physician must decide how to treat a case and the series of treatments to apply. Invasive mycoses may be treated at some point by antifungal agents such as those examples in Table 4.1 (Chapter 4). Shoemaker and House (2005) report success in use of cholestyramine therapy, a polymeric nondegraded oral agent that circulates and binds toxins. It has anion-binding capacity and is said to interrupt the bile carrying capacity of toxins thereby removing them from the system by excretion. Vesper and Vesper (2004) note that a critical mold contaminant is a hemolysin that is proteinaceous. It binds to red cell walls and creates pores causing the red cells to leak (hemolysis). Bacterial hemolysins are said to act more quickly. However, fungal hemolysins, though slower, damage red cells, and in the extreme case, can lead to anemia. The fungi lyse red cells, possibly to access iron needed for their growth and sustenance.

1.5 LEGAL IMPACT

Many states have enacted statutes that cover training and procedures involving mold infestation. None seem to impact regulating construction or quality in housing. OSHA regulates workplace standards. However, this does seem to not apply to residential homes. Also, the concentration standards for mold are not set which may hamper enforcement. There are requirements to check for water behind walls. There may not be any enforcement to check for the actual presence of mold. Nor are there regulations that enforce construction standards by region to optimally prevent water intrusion, the primary incipient cause of mold.

There are cases cited where builders and others have been found negligent. However, proving such cases in court may be difficult unless one can financially afford to support a lawsuit (Mazzulo, 2002; Mummulo, 2009; Powell, 2002; Smith, 2002;

Whited, 2008). Some of these cases are high profile with large awards (Alpert, 2002; Carmody, 2001; Hendrix, 2001). Proving illness as a consequence of unknown mold infestation can be difficult. However, a mold case cited in Texas resulted in a very large award—$32 million (Carmody, 2001). One in San Diego, California, did as well. Two high-profile celebrities, the late Ed McMahon of television fame and Erin Brokovich, won large lawsuits as a consequence of proving mold infestation and harm. Another in Las Vegas concerns a case where a proper metal shield covering plastic water pipe was not in place (G. Goldhammer, personal communication, 2008). An installer hammered drywall into place, unknowingly puncturing the pipe. The nail in place sealed the pipe. However, when the nail rusted away, a water leak resulted, small enough not to be detected, but large enough to cause mold formation, delaying sale of a house and resulting in $25,000 of repair damage. Fortunately, for the individual involved, the developer or insurance paid for the damage. However, this delayed their home sale and move for several months, inconveniencing the parties involved. Not making sure that such a metal protection is in place over a plastic pipe is poor code or possibly negligence.

In another case in California, polybutadiene pipe was used as part of a water system (A. Taub, personal communication, 2009). The joint adhesive failed after several years. The contractor trusted the pipe and adhesive. No one was aware at the time that this failure can occur. Several dwellings were affected. In this case, the fault may lie with the supplier for inadequate testing of its product. Or, it may be that no one is judged to be at fault during litigation. So, if the insurance does not cover the matter (i.e., mold infestation as well as water damage), then the consumer might be stuck with the bill. Again, when one buys a home, it is absolutely necessary to examine these items ad nauseum to protect oneself. Reading this book to get ideas may be a recipe for protection against these and other seemingly unforeseen circumstances.

In this uncertain environment, with the legal and regulatory area not well developed, and insurance possibly not available or protective, the individually affected consumers, absent all other remedies, should research cases and contact law firms that have been successful in lawsuits, if that is their course of action. They can seek to retain such parties or referrals to parties that may help them. This is the best alternative if the only option is a legal option.

REFERENCES

Alpert, D. 2002. Ed McMahon: Death mold killed my dog. ABCNews.com, April 11.

Carmody, K. 2001. Court finds insurer at fault in mold case. *Austin American-Statesman,* June 2.

Conyers, J. 2003. United States Toxic Mold Safety and Protection Act of 2003 ("The Melina Bill") Official Title: To Amend the Toxic Substances Control Act, the Internal Revenue Code of 1986, the Public Buildings Act of 1959, to Protect Human Health from Toxic Mold, and for Other Purposes. H.R. 1268, United States Congress, House of Representatives.

Department of Housing and Urban Development. 2008. Lead Technical Studies and Healthy Homes Technical Studies Programs (FR-5200-N-07).

Goldstein, W., et al. 2008. Sick Building Syndrome. Proceedings of the Annual Meeting of the American Academy of Forensic Sciences (February), Washington, DC.

Gutarowska, B., and Piotrowska, M. 2007. Methods of mycological analysis in buildings. *Building and Environment* 42:1843–1850.

Hendrix, A. 2001. Erin Brockovich crusades against mold. *San Francisco Chronicle*, March 8.

Institute of Medicine. 2004. Damp indoor spaces and health. Washington, DC: The National Academies Press.

Kercsmar, C., Dearborn, D. G., Schluchter, M., Lintong, X., Kirchner, H. L., Sobolewski, J., Greenberg, S. J., Vesper, S. J., and Allan, T. 2006. Reduction in asthma morbidity in children as a result of home remediation aimed at moisture sources. *Environmental Health Perspectives* 114(8) (March):1574–1580.

Mazzulo, L. 2002. "Mold is gold" when building owners file lawsuits. *Wichita Bus. J.*, July 22.

Mudarri, D., and Fisk, W. J. 2007. Public health and economic impact of dampness and mold. *Indoor Air* 17: 226–235.

Mummolo, J. 2009. Sick house, suffering family; Mold leads to major medical problems, legal battle in Loudoun. *The Washington Post*, Suburban Edition, February 1.

Occupational Safety and Health Administration (OSHA). 1970. OSH Act of 1970. 29USC654, Sections 5(1) and 5(2).

Occupational Safety and Health Administration (OSHA). 2008. A brief guide to mold in the workplace. Information Bulletin SHIB 03-10-10, December 2.

Powell, C. 2002. Toxic mold breeds lawsuits. *The Beacon Journal,* October 27.

President's Task Force on Health Risks and Safety Risks to Children. 2000. Asthma and the environment: A strategy to protect children (May).

Sedlbauer, K. 2002a. Prediction of mould fungal formation on the surfaces of and inside building components. Doctoral dissertation, Fraunhofer Institute for Building Physics, Stuttgart, Germany.

Sherman, W. 2007. Couple spent $1.4M on a home but can't get mold out. *New York Daily News*, October 8.

Shoemaker, R. C., and House, D. E. 2005. A time-series study of sick building syndrome: Chronic biotoxin-associated illness from exposure to water-damaged buildings. *Neurotoxicology and Teratology*, 27, 29–46.

Smith, H. 2002. Breaking the mold: As litigation rises, homebuilders working to stamp out spore of the moment. *Las Vegas Review Journal*, September 29.

U.S. Environmental Protection Agency. 2008. Indoor air quality. February 20. www.epa.gov/iaq/pubs/sbs.html.

Vesper, S. J., and Vesper, M. J. 2004. Possible role of fungal hemolysins in sick building syndrome. *Advances in Applied Microbiology* 55:191–213.

Whited, J. 2008. Creeping problem. June 4. LasVegasCityLife.com.

2 Epidemiology and Health Effects in Moisture-Damaged Damp Buildings

Jean Cox-Ganser, PhD
Ju-Hyeong Park, ScD
Richard Kanwal, MD, MPH

CONTENTS

2.1 INTRODUCTION

In recent years there has been much public interest in the health effects of living, attending school, or working in damp buildings with accompanying microbial contamination (especially mold contamination). At the present time, scientific knowledge is not adequate to estimate dose–response relationships between dampness-related exposures and any of the different types of health outcomes suspected to be problems. In fact, the causal agents themselves and effects of being exposed to mixtures of contaminants are not yet understood. Nonetheless, epidemiologic studies have gathered evidence that damp indoor environments are a threat to the health of occupants. A broad range of building-related symptoms and illnesses have gained attention including headache; irritation of eyes, nose, and throat; lack of concentration; rhinitis; lower respiratory symptoms; asthma exacerbation and onset; hypersensitivity pneumonitis; sarcoidosis; infections; nausea; and neurologic effects. Some of these health outcomes have been more thoroughly studied than others; the least studied are the possible health effects of fungal toxins (mycotoxins).

2.2 BACKGROUND

Public interest in indoor environmental quality arose in the 1970s when sick building syndrome (SBS) was first described during a period when changes in building environments and their ventilation had occurred in response to the energy crisis in the United States. Office workers had symptoms of headache, mucous membrane irritation, and difficulty concentrating while in their work buildings, which improved shortly after leaving the buildings. At that time environmental conditions that could lead to such symptoms were not at all understood and widespread complaints, in particular those concerning buildings, were often thought due to mass psychogenic illness. European epidemiologic studies began to establish building-related risk factors for the common symptoms of mucous membrane irritation and discomfort. These included air conditioning, outdoor air ventilation rates, ventilation maintenance practices, housekeeping, and the extent of fleecy materials and open shelving with paper in the indoor environment.

Understanding of these risk factors led to changes in recommendations for ventilation rates and maintenance, and the complaints of nonspecific SBS symptoms alone decreased in the 1990s. Since then the proportion of building-related chest symptoms have increased, at least as reflected in the number of worker requests for health hazard evaluations (HHE) for indoor air quality concerns submitted to the U.S. National Institute for Occupational Safety and Health (NIOSH; 2009). These chest symptoms often arise in buildings with a history of dampness. Well-publicized cases of health effects in relation to water damaged homes with mold contamination also led to increased public health awareness of these issues.

Physicians may be skeptical or simply unaware of the evidence supporting the role of dampness or mold in adverse health effects. A recent editorial in a British journal maintained that many European respiratory physicians "arc reluctant to consider moulds as important in patients with respiratory symptoms, rarely make specific enquiry, and almost never make attempts to reduce mould exposure" (Woodcock, 2007).

Dampness-related exposures in buildings can be highly complex and will vary from building to building and at different locations within a building. Moisture allows growth of bacteria and fungi that usually are already present in building materials or have been brought in from the outdoors by building occupants or outdoor air. Building occupants may be exposed to structural components of these microorganisms (for example, spores and fungal fragments) and to specific substances they may produce; the potential exposures will vary depending on the species of microorganisms that are present and on environmental conditions. Moisture also provides a favorable environment for organisms such as cockroaches and dust mites (exposures related to these organisms have been associated with asthma). Exposures to chemical compounds released by water-damaged building materials (for example, dry wall, vinyl flooring, and carpets) may also occur. In the United States, existing evidence suggests that the prevalence of dampness problems is quite substantial. A survey of 100 U.S. office buildings found that 43 percent had ongoing leaks causing water damage and 85 percent had past water damage (Environmental Protection Agency, 2009). The population-weighted average prevalence of dampness or mold from seven studies on U.S. homes was 47 percent (Mudarri and Fisk, 2007). A Government Accounting Office (GAO) report on school buildings across the United States in 1995, found that about 30 percent of schools needed repairs for roofs, external walls, windows, doors, and plumbing. More recently, a report on Boston public schools indicated that for the 2004–2005 school year, 85 percent of the schools had leaks or water stains, while 63 percent had visible mold growth (Graham et al., 2009). A number of European studies have reported dampness or mold occurring in about 10 to 50 percent of homes (Antova et al., 2008).

2.3 SCOPE AND LIMITATIONS

This chapter concentrates on epidemiologic evidence relating to health effects associated with damp indoor environments, such as homes, schools, and office buildings, with emphasis on research relating to mold contamination. Also included is some discussion of clinical studies and laboratory experiments where this may give evidence of possible exposure effects. Not discussed are the health effects of exposure to animal allergens.

2.4 FINDINGS OF RECENT REVIEWS

In 2004 and 2009, two major reviews were published relating to health effects and damp indoor environments. These were the Institute of Medicine report (IOM, 2004) and the World Health Organization guidelines for dampness and mold (WHO, 2009).

The IOM report covered pertinent literature published up to late 2003 on a wide range of health effects including nasal and throat symptoms, cough, wheezing, shortness of breath, lower respiratory illness in otherwise healthy children and adults, asthma exacerbation, asthma development, airflow obstruction, mucus membrane irritation syndrome, chronic obstructive pulmonary disease, inhalation fevers, skin symptoms, gastrointestinal tract problems, fatigue, neuropsychiatric symptoms, cancer, reproductive effects, rheumatologic and other immune diseases, and acute idiopathic pulmonary hemorrhage in infants.

Major findings were that sufficient evidence existed for associating the presence of mold or other agents in damp buildings with nasal and throat symptoms, cough, wheezing, asthma exacerbations, and hypersensitivity pneumonitis. The IOM committee concluded that limited or suggestive evidence existed for associating exposure to damp indoor environments with shortness of breath, asthma development, and, in otherwise healthy children, lower respiratory disease. There was insufficient or inadequate evidence to determine if associations exist for the other health effects listed. The committee noted that immunocompromised individuals are at risk for fungal infections. They concluded that excessive indoor dampness is a public health problem and that prevention or reduction of this condition should be a public health goal (IOM, 2004).

The WHO guidelines covered literature published up to July 2007 on upper respiratory tract symptoms, cough, wheeze, shortness of breath, asthma exacerbation, asthma development, current asthma, respiratory infections, bronchitis, wheezing, allergic rhinitis, and allergy. It did not cover effects related to skin, eyes, fatigue, nausea, headache, insomnia, mucous membrane irritation, or SBS (WHO, 2009).

In the chapter on health effects, the authors of the WHO guidelines concluded that there is sufficient epidemiological evidence of an association between indoor dampness-related factors and asthma development, asthma exacerbation, current asthma, respiratory infections, upper respiratory tract symptoms, cough, wheezing, and shortness of breath. Updated findings comparing the WHO guidelines and IOM report are that there is now sufficient evidence of an association between indoor-dampness-related agents and the development of asthma, current asthma, shortness of breath, and respiratory infections. There is now limited or suggestive evidence for bronchitis and allergic rhinitis being related to dampness-related exposures. There is clinical evidence that exposure to mold and other dampness-related microbial agents increase the risk for hypersensitivity pneumonitis, chronic rhinosinusitis, and allergic fungal sinusitis. Importantly, the WHO guidelines note that although atopy and allergy increase susceptibility to dampness-related health effects, such health effects are also found in nonatopic building occupants.

In agreement with the recent WHO guidelines, a 2008 review of literature published since the IOM report concluded that the recent epidemiological studies provided additional evidence for an association between damp indoor environments and asthma development and shortness of breath. Importantly, this review discusses evidence that remediation of damp, water-damaged buildings reduces respiratory health effects, but that lower respiratory symptoms may take some time to resolve and dampness-related asthma may not completely resolve (Sahakian et al., 2008)

In 2008, the American Industrial Hygiene Association (AIHA) published a book titled *Recognition, Evaluation, and Control of Indoor Mold*, in which the first chapter is devoted to health concerns and emphasizes the importance that health complaints by building occupants plays in risk management (Prezant et al., 2008). Many practicing industrial hygienists will use this influential book, which also discusses building evaluation for moisture, identifying mold damage, sample collection and evaluation, and the process of remediation and control.

2.5 SICK BUILDING SYNDROME AND OTHER BUILDING-RELATED SYMPTOMS

Although we discuss building-related symptoms separately from building-related illnesses, it must be realized that symptoms may be evidence of subclinical or undiagnosed illness. Studies of the health effects of damp indoor environments often report on both building-related symptoms and diagnosed illnesses reflecting the underlying fact that these health effects occur together in populations of building occupants.

A number of studies over the years have indicated that increasing building ventilation will lower SBS symptom prevalence. An analysis that combined data from such studies found that increasing the ventilation rate from 10 to 25 liters per second per person was associated with about a 29 percent decrease in SBS symptoms (Fisk et al., 2009). Poorly maintained humidification systems, cooling coils, and drainage pans in office buildings are also associated with increased prevalence of SBS symptoms, such as upper respiratory symptoms, fatigue/difficulty concentrating, skin symptoms, eye symptoms, and headache; with odds ratios of about 1.5 (Mendell et al., 2008).

A recent update on risk factors for SBS symptoms noted building dampness and mold, the inflammatory potential of indoor particles, and psychosocial factors (Norback, 2009). Also remarked upon was the increasing recognition of the importance of SBS symptoms in home environments, especially in Asia, with recognition of chemical exposures and mold as risk factors, and the coining of the term *sick house syndrome*. A unique large cohort study on a random sample of 1,000 adults in Sweden looked at changes in SBS symptoms during a 10-year follow-up period in relation to changes in indoor exposures at home, including dampness and mold (Sahlberg et al., 2009). The study found increased risks in relation to dampness or mold in the home during the follow-up period for onset of skin symptoms (risk ratio [RR] 2.3, 1.4–3.9), mucosal symptoms (RR 3.2, 1.7–5.9), and general symptoms (RR 2.2, 1.3–3.7).

An extreme form of dampness and mold problems in buildings occurs after flooding linked to natural disasters. A study of about 550 homes in New Orleans six months after Hurricanes Katrina and Rita, in which one adult per site was interviewed about remediation activities and upper and lower respiratory symptoms, found that respiratory symptom scores increased linearly with exposure to water-damaged homes (Cummings et al., 2008).

Since the mid-1990s, a 20-story office building in the northeastern United States had experienced chronic moisture problems from rainwater leaks through the roof, around windows, and through sliding doors of terraces. The upper floors had suffered the most water damage and mold growth. Many floors had also experienced plumbing leaks that had damaged interior walls (Cox-Ganser et al., 2005). As part of an HHE of the building, NIOSH conducted a questionnaire survey of building workers in 2001; 888 of 1,327 workers (67 percent) participated. The proportion of workers who reported wheezing, nose, or eye symptoms that improved on days off work was 3.4 times higher than expected compared to the U.S. adult population. The proportions of workers reporting frequent wheezing, coughing attacks, chest tightness, and shortness of breath in the last four weeks that improved when away from the building were 2.7 to 4.6 times higher than expected in comparison to data from a study of office workers in buildings with no known indoor environmental problems (Cox-Ganser et al., 2005). In this population, associations were found between building-floor mean levels of fungi and endotoxins in settled dust and building-related upper and lower respiratory symptoms (Park et al., 2006).

Two large, recently published studies on respiratory symptoms in children have emphasized home dampness and mold as risk factors. The prevalence of respiratory and allergic symptoms in association with home exposures to environmental tobacco smoke (ETS) and dampness or mold was investigated in 4,122 Sardinian children from 29 primary schools. The study had a high response rate of 84.4 percent. Study results confirmed ETS as a risk factor and also found that exposure to dampness or mold in the home was associated with increased prevalence of current wheezing, persistent cough or phlegm, and current rhinoconjunctivitis (Pirastu et al., 2009). Data on 57,161 children (6 to 12 years old) from 12 cross-sectional studies carried out in Russia, North America, and 10 countries in Eastern and Western Europe were pooled and analyzed for associations between visible mold in homes and eight respiratory and allergic symptoms (Antova et al., 2008). Consistent, positive associations between exposure to home dampness or mold and children's respiratory health were found with odds ratios from 1.3 to 1.5. Despite these findings, only 13 percent of parents in the studies from Bulgaria, Czech Republic, Hungary, Poland, and Slovakia who took part in a risk perception survey thought home dampness or mold was a cause of the children's breathing problems.

2.6 BUILDING-RELATED ILLNESS

2.6.1 Rhinitis and Sinusitis

Nasal stuffiness, sneezing, and a runny or itchy nose characterize rhinitis. Workers in damp buildings who experience these symptoms while at work with improvement when away from the building may have rhinitis due to a dampness-related exposure. The rhinitis may be due to an allergic response (possibly to a mold species) or an irritant response. Sinusitis (inflammation of the paranasal sinuses) can cause symptoms similar to those of rhinitis or a cold. Sinusitis is usually caused by viruses or bacteria and less often by fungi. Inhalation of irritant substances can also be a cause. Allergic rhinosinusitis is one of the most common chronic diseases, and its worldwide prevalence is increasing. Estimated prevalence ranges between 10 to 50 percent (Bellanti and Wallerstedt, 2000). The prevalence of work-related rhinitis in people with occupational asthma has been estimated between 76 to 92 percent (Siracusa et al., 2000). In addition, there is a substantial body of information linking asthma risk and exacerbation to upper respiratory tract symptoms (Passalacqua et al., 2004). An estimated annual cost of rhinosinusitis was \$2 billion to \$5 billon in the United States in 2003 (Reed et al., 2004).

Epidemiologic studies of upper respiratory symptoms consistent with rhinitis and sinusitis have been discussed in the previous section on building-related symptoms. The 2009 WHO guidelines determined that there was limited or suggestive evidence of an association between dampness-related exposures and allergic rhinitis based on six studies in children (WHO, 2009).

2.6.2 Asthma

Asthma is a chronic disease of the lung airways characterized by inflammation and episodes of airways obstruction. Some individuals develop asthma due to an allergic response, while in others the disease mechanism is nonallergic. Symptoms associated with airways obstruction include wheezing, chest tightness, shortness of breath, and cough. The airways obstruction is at least partially reversible with medications (for example, inhaled bronchodilators such as albuterol) or may resolve spontaneously with time. Lung function testing with spirometry may reveal airflow obstruction that responds to administration of a bronchodilator. A methacholine challenge test may reveal evidence of airways hyperreactivity (sensitive airways, as indicated by a significant decrease in airflow after inhalation of a low concentration of methacholine). Methacholine challenge (or another nonspecific bronchoprovocation test) may be useful to establish the diagnosis in individuals with asthma symptoms who have a normal spirometry test. If asthma is not adequately treated, chronic inflammation over time may progress to severe asthma with airways obstruction that is no longer fully responsive to bronchodilator medications (fixed airways obstruction).

Asthma is a fairly common disease. The prevalence of currently active asthma in adults in the United States was approximately 6 percent in 2002; the adult lifetime prevalence (asthma at any point in a person's life) was approximately 10 percent (Apter and Weiss, 2008). It has been estimated that 21 percent of current U.S. asthma cases are attributable to dampness and mold, which translates to 4.6 million of 21.8 million asthma cases attributable to dampness and mold (Mudarri and Fisk, 2007). This study also estimated the annual cost of asthma attributable to dampness and mold exposure at $3.5 billion.

In the United States, NIOSH investigated respiratory symptoms and asthma in relation to damp indoor environments in employees of two hospitals. Over a two-year period, 6 of 53 workers (11 percent) located on the top floor of one of these hospitals developed asthma. The hospital had experienced multiple episodes of significant roof and window leaks during heavy rains over several years. Walls and ceilings showed mold growth. Five of the six affected workers had no previous asthma history, while one worker had a history of childhood asthma but had not had any asthma symptoms in 20 years. All six workers reported asthma symptoms that improved away from work and had evidence of bronchial hyperresponsiveness on methacholine challenge testing. Blood analyses showed no allergic response to latex or common environmental allergens (e.g., house dust mites, grass, cat, and dog) in any of the affected workers. In four of the six affected workers, serial peak flow measurements showed a work-related pattern (declines in peak flow during work days and improvement when off work); of the other two affected workers, one had a mixed pattern (drops in peak flow at work and at home), and the other showed some declines in peak flow over the work week (Cox-Ganser et al., 2009).

A survey of respiratory health in 1,171 of 1,834 employees working in the two hospitals combined with environmental assessment for dampness and biological contaminants found that posthire asthma and work-related lower respiratory symptoms were positively associated with dampness. Furthermore work-related lower respiratory symptoms showed monotonically increasing odds ratios with ergosterol, a marker of fungal biomass.

In the water-damaged office-building study in the 1990s discussed earlier in the section on building-related symptoms, the prevalence of current asthma among participating workers was 2.4 times higher than expected in comparison to the general U.S. adult population data from the Third National Health and Nutrition Examination Survey, and 1.5 times higher than expected in comparison to state-specific data from the Behavioral Risk Factor Surveillance System; workers reported asthma onset at estimated rates that were 7.5 times higher after starting work in the building than before starting work in the building (Cox-Ganser et al., 2005). Further study linking health effects to estimates of microbial contamination in settled dust found respiratory illnesses showed significant linear exposure–response relationships to total culturable fungi (interquartile range odds ratios [IQR-OR] 1.4–1.7), hydrophilic fungi (IQR-OR 1.4–2.2), and ergosterol (IQR-OR 1.5–1.6) in floor and chair dusts. Postoccupancy physician-diagnosed asthma was most strongly associated with exposure to hydrophilic fungi (fungi with a water-activity requirement of 0.9 or more; Park et al., 2008).

Two compelling recent population-based incident case-control studies in Finland, one on adults and one on children, have shown that water damage and mold in the workplace and at home is associated with the onset of asthma. The study on adults aged 21 to 63, compared 521 people with newly diagnosed asthma to 932 randomly selected controls without asthma. The presence of workplace wall-to-wall carpeting and workplace mold independently increased the risk of new-onset asthma (Jaakkola et al., 2006). The study of children aged 1 to 7 years old compared 121 asthma cases and 241 controls and found odds ratios of 2.6 for visible mold in the main living area, and 2.2 for water damage in the main living area (Pekkanen et al., 2007).

2.6.3 Hypersensitivity Pneumonitis (HP)

Hypersensitivity pneumonitis (HP) is a serious lung disease in which an individual's immune system responds to repeated inhalation of organic matter (material from living things such as plants, animals, bacteria, or fungi) or other sensitizing agents. Dozens of different fungi, bacteria, animal proteins, plants, and chemicals are known causes of HP (Patel et al., 2001). Examples of occupations in which HP is known to occur include farmers exposed to dust from moldy hay and machinists exposed to metalworking fluid mists. There have been many reports in the scientific literature of individuals who have developed HP while working in damp office buildings and schools, or living in homes with evidence of moisture damage and mold (Apostolakos et al., 2001; Cox-Ganser et al., 2005; Hoffman et al., 1993; Thorn et al., 1996; Weltermann et al., 1998).

Two symptom patterns exist in HP. Some individuals experience episodic shortness of breath and flulike symptoms, including cough, muscle aches, chills, fever, sweating, and fatigue (acute disease). These symptoms start within hours of exposure and last for one to three days. Other individuals do not develop flulike symptoms; instead they develop gradual and progressive shortness of breath and cough, often accompanied by weight loss. Physicians often do not consider the possibility of HP when evaluating a patient. An individual with rapid onset of flulike symptoms and shortness of breath due to HP might be diagnosed with a respiratory infection. In HP cases that are caused by occupational exposures, the first sign that the illness is due to exposures at work may be that a worker's symptoms and medical tests improve during a period of time away from work and then worsen on return to work.

The main treatment for HP is cessation of further exposure to the causative agent. Acute disease may resolve completely with exposure cessation; corticosteroid medications may shorten disease duration. With long-term exposure, the disease may not improve (or may continue to worsen and progress to permanent lung scarring) even after exposure ceases.

2.6.4 Sarcoidosis

Another disease that can have similar respiratory symptoms and medical evaluation findings as occur in HP is sarcoidosis. In sarcoidosis, the lungs and many other organ systems can develop granulomas and have abnormal function. There are no definitive medical tests that establish a diagnosis of sarcoidosis. It is a diagnosis of exclusion once other known causes of granulomatous lung disease (e.g., HP, chronic beryllium disease, certain infections) have been ruled out (Newman et al., 1997). An individual diagnosed as having sarcoidosis but who actually has HP from an occupational exposure is at risk for progression to severe irreversible lung scarring from continued exposure in the workplace (Forst and Abraham, 1993). The cause of sarcoidosis is unknown but the occurrence of clusters of cases suggests that environmental factors may be responsible for some cases (Newman et al., 1997). A study of over 700 sarcoidosis cases found associations with work in areas with musty odors and with occupational exposure to insecticides (Newman et al., 2004). Clusters of sarcoidosis cases have been reported among workers in buildings with persistent dampness (Cox-Ganser et al., 2005; Laney et al., 2008).

2.6.5 Possible Mycotoxin-Related Health Effects

In late 2008, the GAO reported to the Committee of Health, Education, Labor, and Pensions, that federal agencies should better coordinate their research on health effects of indoor mold and deliver more consistent guidance to the public (GAO, 2008). One of the observations was that federal and federally sponsored research had not extensively studied mycotoxins in relation to the indoor environment including acute pulmonary hemorrhage in infants, dose– or exposure–response relations, techniques to quantify mycotoxins in tissue and environmental samples, and health effects from long-term exposure to mycotoxins.

The WHO guidelines discussed mycotoxins in the section on clinical aspects of health effects and stated that the evidence that mycotoxins play a role in health problems related to damp indoor environments is very weak (WHO, 2009). An indication of biological effects of indoor exposure to mycotoxins comes from a small study on people exposed to mycotoxins from *Stachybotrys chartarum* in their homes. Albumin adducts were found to form in their blood, as found in rats exposed experimentally to these mycotoxins (Yike et al., 2006).

Mycotoxins are secondary metabolites produced by fungi for survival from environmental challenges such as nutrient starvation or the presence of competing organisms. More than 400 mycotoxins in at least 21 different groups produced by more than 350 fungi have been isolated thus far (Hussein and Brasel, 2001; Kuhn and Ghannoum, 2003), and many mycotoxins have the potential of being toxic to human beings and animals even at very low concentrations.

In vitro and animal studies suggest that some mycotoxins can produce immunotoxicity by either suppressing or stimulating immune systems (IOM, 2004). As an immune suppressor, mycotoxins can inhibit protein syntheses or cell proliferation. Immune stimulating effects can lead to hypersensitivity reactions (Sharma, 1993). Research using human peripheral blood mononuclear cells showed that exposure to mycotoxins polarized helper T cells toward the Th2 phenotype by inhibiting Th1 cells producing INF-γ (Wichmann et al., 2002). The authors suggested that exposure to mycotoxins may increase the risk for development of allergies. In animal studies, neurotoxic effects of mycotoxins exposure have been also reported. Occupants of water-damaged buildings sometimes report fatigue, headache, memory loss, and depression; however, it is not yet known whether these central nervous system symptoms are due to neurotoxic effects of mycotoxin exposure (IOM, 2004).

One reason for the lack of epidemiologic studies on mycotoxin exposure in indoor environments has been the difficulty in measuring low levels of airborne mycotoxins (IOM, 2004). Recent interest in quantifying indoor mycotoxins among researchers has driven an improvement in measurement methods. Gas chromatography-tandem mass spectrometry, high performance liquid chromatography-electronspray tandem mass spectrometry, liquid chromatography-mass spectrometry, Fourier transform infrared spectroscopy, and nuclear magnetic resonance have been applied for determination and quantification of mycotoxins in samples (Bloom et al., 2007; de la Campa et al., 2007; Larsson, 2008; Slack et al., 2009). Macrocyclic trichothecene-specific enzyme-linked immunosorbent assays have also been developed to detect mycotoxins produced by *Stachybotrys chartarum* (Brasel, Douglas, et al., 2005; Brasel, Martin, et al., 2005). Improved detection techniques have led to reports that mycotoxins can be found in bulk materials, settled dusts, intact spores, and fungal fragments from water-damaged buildings (Bloom et al., 2009; Brasel, Douglas, et al., 2005; Charpin-Kadouch et al., 2006).

REFERENCES

Antova, T., Pattenden, S., Brunekreef, B., Heinrich, J., Rudnai, P., Forastiere, F., Luttmann-Gibson, H., et al. 2008. Exposure to indoor mould and children's respiratory health in the PATY study. *J Epidemiol Community Health* 62(8): 708–714.

Apostolakos, M. J., Rossmoore, H., and Beckett, W. S., 2001. Hypersensitivity pneumonitis from ordinary residential exposures. *Environ Health Perspect* 109(9): 979–981.

Apter, A. J., and Weiss, S. T. 2008. Asthma: Epidemiology. In *Fishman's pulmonary diseases and disorders*, 4th ed, Chapter 46, ed. A.P. Fishman. New York: McGraw-Hill.

Bellanti, J. A., and Wallerstedt, D. B. 2000. Allergic rhinitis update: Epidemiology and natural history. *Allergy Asthma Proc* 21(6): 367–370.

Bloom, E., Bal, K., Nyman, E., and Larsson, L. 2007. Mass spectrometry-based strategy for direct detection and quantification of some mycotoxins produced by *Stachybotrys* and *Aspergillus* spp. in indoor environments. *Appl Environ Microbiol* 73(13): 4211–4217.

Bloom, E., Grimsley, L. F., Pehrson, C., Lewis, J., and Larsson, L. 2009. Molds and mycotoxins in dust from water-damaged homes in New Orleans after hurricane Katrina. *Indoor Air* 19(2): 153–158.

Brasel, T. L., Douglas, D. R., Wilson, S. C., and Straus, D. C., 2005. Detection of airborne *Stachybotrys chartarum* macrocyclic trichothecene mycotoxins on particulates smaller than conidia. *Appl Environ Microbiol* 71(1): 114–122.

Brasel, T. L., Martin, J. M., Carriker, C. G., Wilson, S. C., and Straus, D. C. 2005. Detection of airborne *Stachybotrys chartarum* macrocyclic trichothecene mycotoxins in the indoor environment. *Appl Environ Microbiol* 71(11): 7376–7388.

Charpin-Kadouch, C., Maurel, G., Felipo, R., Queralt, J., Ramadour, M., Dumon, H., Garans, M., Botta, A., and Charpin, D. 2006. Mycotoxin identification in moldy dwellings. *J Appl Toxicol* 26(6): 475–479.

Cox-Ganser, J. M., Rao, C. Y., Park, J.-H., Schumpert, J. C., and Kreiss, K. 2009. Asthma and respiratory symptoms in hospital workers related to dampness and biological contaminants. *Indoor Air* 19(4): 280–290.

Cox-Ganser, J. M., White, S. K., Jones, R., Hilsbos, K., Storey, E., Enright, P. L., Rao, C. Y., and Kreiss, K. 2005. Respiratory morbidity in office workers in a water-damaged building. *Environ Health Perspect* 113(4): 485–490.

Cummings, K. J., Cox-Ganser, J., Riggs, M. A., Edwards, N., Hobbs, G. R., and Kreiss, K. 2008. Health effects of exposure to water-damaged New Orleans homes six months after Hurricanes Katrina and Rita. *Am J Public Health* 98(5): 869–875.

de la Campa, R., Seifert, K., and Miller, J. D. 2007. Toxins from stains of *Penicillium chrysogenum* isolated from building and other sources. *Mycopathologia* 163(3): 161–168.

Environmental Protection Agency (EPA). 2009. BASE study. http://www.epa.gov/iaq/base/summarized_data.html (accessed November 23, 2009).

Fisk, W. J., Mirer, A. G., and Mendell, M. J. 2009. Quantitative relationship of sick building syndrome symptoms with ventilation rates. *Indoor Air* 19(2): 159–165.

Forst, L., and Abraham, J. 1993. Hypersensitivity pneumonitis presenting as sarcoidosis. *Br J Ind Med* 50(6): 497–500.

General Accounting Office (GAO). 1995. *School facilities: Condition of America's Schools* (GAO/HEHS-95-61). Washington, DC: U.S. General Accounting Office.

General Accounting Office (GAO). 2008. *Indoor Mold* [online]. //www.gao.gov/new-items/d08980.pdf (accessed November 23, 2009).

Graham, T., Zotter, J., and Camacho, M. 2009. Who's sick at school: Linking poor school conditions and health disparities for Boston's children. *New Solutions* 19(3): 355–364.

Hoffman, R. E., Wood, R. C., and Kreiss, K. 1993. Building-related asthma in Denver office workers. *Am J Public Health* 83(1): 89–93.

Hussein, H. S., and Brasel, J. M. 2001. Toxicity, metabolism, and impact of mycotoxins on humans and animals. *Toxicology* 167(2): 101–134.

Institute of Medicine (IOM). 2004. *Damp Indoor Spaces and Health.* Washington, DC: The National Academies Press.

Jaakkola, J. K., Ieromnimon, A., and Jaakkola, M.S. 2006. Interior surface materials and asthma in adults: A population-based incident case-control study. *Am J Epidemiol*, 164(8): 742–749.

Kuhn, D. M., and Ghannoum, M. A. 2003. Indoor mold, toxigenic fungi, and *Stachybotrys chartarum*: Infectious disease perspective. *Clin Microbiol Rev* 16(1): 144–172.

Laney, A. S., Cragin, L. A., Blevins, L. Z., Sumner, A. D., Cox-Ganser, J. M., Kreiss, K., Moffatt, S. G., and Lohff, C. J. 2008. Sarcoidosis, asthma, and asthma-like symptoms among occupants of a historically water-damaged office building. *Indoor Air* 19(1): 83–90.

Larsson, L. 2008. Use of mass spectrometry for determining microbial toxins in indoor environments. *J Environ Monit*, 10(3): 301–410.

Mendell, M. J., Lei-Gomez, Q., Mirer, A. G., Seppanen, O., and Brunner, G. 2008. Risk factors in heating, ventilating, and air-conditioning systems for occupant symptoms in US office buildings: The US EPA BASE study. *Indoor Air* 18(4): 301–316.

Mudarri, D., and Fisk, W. J. 2007. Public health and economic impact of dampness and mold. *Indoor Air* 17(3): 226–235.

National Institute for Occupational Safety and Health (NIOSH). 2009. *Health Hazard Evaluation Program* [online]. http://www.cdc.gov/niosh/hhe/HHEprogram.html (accessed November 23, 2009).

Newman, L. S., Rose, C. S., and Maier, L. A. 1997. Sarcoidosis. *N Engl J Med*, 336(17):1224–1234.

Newman, L. S., Rose, C. S., Bresnitz, E. A., Rossman, M. D., Barnard, J., Frederick, M., Terrin, M. L., et al. 2004. A case control etiologic study of sarcoidosis: Environmental and occupational risk factors. *Am J Respir Crit Care Med* 170 (12): 1324–1330.

Norback, D. 2009. An update on sick building syndrome. *Curr Opin Allergy Clin Immunol* 9(1): 55–59.

Park, J.-H., Cox-Ganser, J., Rao, C., and Kreiss, K. 2006. Fungal and endotoxin measurements in dust associated with respiratory symptoms in a water-damaged office building. *Indoor Air* 16(3): 192–203.

Park, J.-H., Cox-Ganser, J. M., Kreiss, K., White, S. K., and Rao, C. Y. 2008. Hydrophilic fungi and ergosterol associated with respiratory illness in a water-damaged building. *Environ Health Perspect* 116(1): 45–50.

Passalacqua, G., Ciprandi, G., Pasquali, M., Guerra, L., and Canonica, G. W. 2004. An update on the asthma-rhinitis link. *Curr Opin Allergy Clin Immunol* 4(3): 177–183.

Patel, A., Ryu, J., and Reed, C. 2001. Hypersensitivity pneumonitis: Current concepts and future questions. *J Allergy Clin Immunol* 108(5): 661–670.

Pekkanen, J., Hyvärinen, A., Haverinen-Shaughnessy, U., Korppi, M., Putus, T., and Nevalainen, A. 2007. Moisture damage and childhood asthma: A population-based incident case-control study. *Eur Respir J* 29(3): 509–515.

Pirastu, R., Bellu, C., Greco, P., Pelosi, U., Pistelli, R., Accetta, G., and Annibale Biggeri, A. 2009. Indoor exposure to environmental tobacco smoke and dampness: Respiratory symptoms in Sardinian children—DRIAS study. *Environ Res* 109(1): 59–65.

Prezant, B., Weekes, D. M., and Miller, J. D. (Eds.). 2008. *Recognition, evaluation, and control of indoor mold.* Fairfax, VA: American Industrial Hygiene Association.

Reed, S. D., Lee. T. A., and McCrory, D. C. 2004. The economic burden of allergic rhinitis: A critical evaluation of the literature. *Pharmacoeconomics* 22(6): 345–361.

Sahakian, N. M., Park, J.-H., and Cox-Ganser, J. M. 2008. Dampness and mold in the indoor environment: Implications for asthma. *Immunol Allergy Clin North Am* 28(3): 485–505.

Sahlberg, B., Wieslander, G., and Norback, D. 2009. Sick building syndrome in relation to domestic exposure in Sweden—A cohort study from 1991 to 2001. *Scand J Public Health*: 1–7.

Sharma, R. P. 1993. Immunotoxicity of mycotoxins. *J Dairy Sc* 76(3): 892–897.

Siracusa, A., Desrosiers, M., and Marabini, A. 2000. Epidemiology of occupational rhinitis: Prevalence, aetiology and determinants. *Clin Exp Allergy* 30(11): 1519–1534.

Slack, G. J., Puniani, E., Frisvad, J. C., Samson, R. A., and Miller, J. D. 2009. Secondary metabolites from *Eurotium* species, *Aspergillus calidoustus* and *A. insuetus* common in Canadian homes with a review of their chemistry and biological activities. *Mycol Res* 113: 480–490.

Thorn, A., Lewne, M., and Belin, L. 1996. Allergic alveolitis in a school environment. *Scan J Work Environ Health* 22(4): 311–314.

Weltermann, B. M., Hodgson, M., Storey, E., DeGraff, A. C., Bracker, A., Groseclose, S., Cole, S. R., Cartter, M., and Phillips, D. 1998. Hypersensitivity pneumonitis: A sentinel event investigation in a wet building. *Am J Ind Med* 34(5): 499–505.

Wichmann, G., Herbarth, O., and Lehmann, I. 2002. The mycotoxins citrinin gliotoxin and patulin affect interferon-γ rather than interleukin-4 production in human blood cells. *Environ Toxicol* 17(3): 211–218.

Woodcock, A. 2007. Moulds and asthma: Time for indoor climate change? *Thorax* 62(9): 745–746.

World Health Organization (WHO). 2009. *WHO Guidelines for Indoor Air Quality: Dampness and Mould.* WHO Regional Office for Europe.

Yike, I., Distler, A. M., Ziady, A. G., and Dearborn, D. G. 2006. Mycotoxin adducts on human serum albumin: Biomarkers of exposure to *Stachybotrys chartarum. Environ Health Perspect* 114(8): 1221–1226.

3 Mold Biology, Molecular Biology, and Genetics

Edward Sobek, PhD

CONTENTS

3.1 BIOLOGY OF INDOOR MOLDS: AN OVERVIEW

There are an estimated 1.5 million fungal species worldwide. However, only about 100,000 or 5 percent of the 1.5 million species have been identified (Mueller and Schmit, 2007). Of the 100,000 known fungi only a small percentage (<0.1%) are commonly associated with the indoor environment (Macher, 1999; Samson et al., 2002; Vesper et al., 2007a; Zorman and Jersek, 2008). The mold species that do occur indoors have adapted to occupy a unique ecological niche that fosters little competition but an abundance of resources.

Indoor molds also comprise a group of fungal species that are pathogens of humans (Rhodes and Brakhage, 2006) and cause a variety of allergy and respiratory related illnesses (Ward et al., 2007). Immune compromised patients are particularly at risk from the pathogenic molds while children are considered a high-risk group when exposed to allergy and asthma inducing molds (Vesper et al., 2007b).

The defining biological characteristics of indoor molds are (1) an asexual life cycle, (2) the abundant production of microscopic spores called conidia, and (3) the ability to colonize and degrade a variety of indoor substrates including wood-based products, glues, polymers, and fabrics.

3.2 CLASSIFICATION OF INDOOR MOLDS

The vast majority of indoor molds belong to the eumycota or true fungi. There are four recognized groups of eumycotan fungi: ascomycetes, zygomycetes, basidiomycetes, and chytrids. Other indoor molds include myxomycetes, which are actually protists not fungi; however, they are still primarily studied by mycologists. To facilitate discussion, the author classifies molds in this chapter into four groups (Table 3.1). They are the ascomycetous molds (Ascos), zygomycetous molds (Zygos), Yeasts, and myxomycetous molds (Myxos). Of these, the Ascos dominate Table 3.1. The yeasts are unicellular, asexual budding molds. The myxomycetous molds are limited to a few species of stalked, sporangia-bearing forms. The Ascos comprise a large group of fungi that incorporate plant pathogens, macrofungi (morels) endophytes, parasites, and symbiants (Hawksworth et al., 1995). Ascos typically have distinct hyphal cross walls that occur at regular intervals. They include species such as *Aspergillus*, *Stachybotrys*, *Penicillium*, and *Cladosporium*. Species of zygomycetes lack regular hyphal cross walls and include species of *Mucor* and *Rhizopus*.

3.3 THE LIFE CYCLE OF INDOOR MOLDS

All indoor environments have some mold spores, with the possible exception of the strictest clean room environment. Studies by the U.S. Environmental Protection Agency (EPA), using DNA detection technology, have significantly advanced the current understanding of the mold burden in homes (Vesper et al., 2007a). By separating molds into two distinct groups the EPA has identified 30 species of water intrusion molds and 10 species of common background molds. The common background molds correlate with indoor hygiene. These molds end up indoors by hitching a ride

TABLE 3.1
Common Indoor Molds Represented In Each Major Taxonomic Group

Ascos	Zygos	Yeasts	Myxos
Acremonium	*Absidia*	*Candida*	*Didymium*
Alternaria	*Mucor*	*Cryptococcus*	*Physarum*
Aspergillus	*Rhizopus*	*Rhodotorula*	*Stemonitis*
Aureobasidium	*Syncephalastrum*	*Sporobolomyces*	
Chaetomium			
Cladosporium			
Curvularia			
Epicoccum			
Eurotium			
Fusarium			
Memnoniella			
Nigrospora			
Paeciliomyces			
Penicillium			
Pithomyces			
Scopulariopsis			
Spegazzinia			
Stachybotrys			
Trichoderma			
Ulocladium			
Wallemia			

on shoes, pets, or on wind currents through open windows and doors. Homes that are consistently cleaned, dusted, and vacuumed have a lower burden of background molds compared to indoor environments that are cleaned infrequently.

The water intrusion mold species are usually absent or in very low concentrations indoors, unless humidity or wet conditions exist. In such cases, the water intrusion molds significantly outnumber the background molds in both concentration and diversity of species present. The important point to note is that the life cycle of problematic indoor molds is initiated with moisture or free water. The background molds are capable of growing in wet indoor environments; however, they are frequently displaced and subdued by the rapid growing and highly competitive water intrusion molds.

3.3.1 Germination

A variety of environmental factors are known to affect fungal spore germination. They include pH, temperature, nutrient availability, and inhibition by competing microbes (Garraway and Evans, 1984). However, for indoor molds, moisture is probably the overarching factor that influences spore germination. Because indoor environments are typically composed of simple, nutrient rich substrates that are mostly void of competitors, most other factors affecting germination are not an issue.

The amount of moisture required for germination is often species specific. Griffin (1994) provides an excellent review of water relations in fungi. In general, some broad generic patterns can be extracted from the fungal literature related to spore germination of indoor molds. For example, it is generally accepted that the ascomycetous molds are adapted to germinate on substrates with low water activity; whereas the zygomycetous molds require water saturated substrates (Webster and Weber, 2007). However, within the ascomycetous molds a wide variation of water activity requirements exists even within a genus. *Stachybotrys chartarum* requires water-saturated conditions, whereas many species of *Penicillium* and *Aspergillus* can grow in overly humid environments. In order to germinate and grow at low water activity, molds have to concentrate the solutes inside spores and hyphae to a concentration greater than the substrate (Griffin, 1994). The higher concentration of solute inside the fungus sets up a gradient that allows water to flow from the substrate to the hyphae.

Spores are extremely adept at water uptake as evident by spore ultrastructure and chemical content. The diverse beta glucan content of spore walls and trehalose sugar inside spores allow significant uptake of water from the substrates and humid air. A few mold species are extreme solute concentrators, such as *Wallemia sebi*, which can germinate at a water activity of 0.69 (Samson et al., 2002). Indoor mold species that can germinate on substrates that have water activities less than 0.75, which include *Wallemia*, *Eurotium*, *Aspergillus*, and *Penicillium* species, are commonly encountered in overly humidified indoor environments (>65% relative humidity). These low water activity molds contribute to the common phenomenon of surface mold that occurs throughout the southeastern United States during late summer and early fall. Articles and clothing kept in unairconditioned spaces such as garages and humid basements are prime locations for surface mold germination and subsequent colonization.

3.3.2 Indeterminate Growth

Most organisms have a preprogrammed size and lifespan that is determined by the underlying genetic configuration of the individual. Humans for example, stop growing in height around age 20 and persist into their late 70s. The molds, however, exhibit the unique characteristic of indeterminate growth. Indeterminate growth is defined as continuous genetic propagation without meiotic recombination. Hence, virtually any portion of the mold's vegetative form or mycelium will grow indefinitely as long as environmental conditions are conducive to growth. In addition, the vegetative form may also serve as a propagative structure. If mycelium is disturbed and redistributed onto a new, growth-conducive substrate, it serves as an extension of the original growing mycelium. In fact, it is a clone of the original fungus and functions under the same genetic restrictions as the mycelium it was cloned from. Likewise, asexual spores or conidia act as individual clones of the original mycelium. Hence, molds do not experience the same restrictions of space and time that most other organisms are confined to. Some mold clones may have originated hundreds or thousands of years ago and traveled around the globe. The development of novel molecular tools may one day allow for the determination of a mold clone's age and geographic origin.

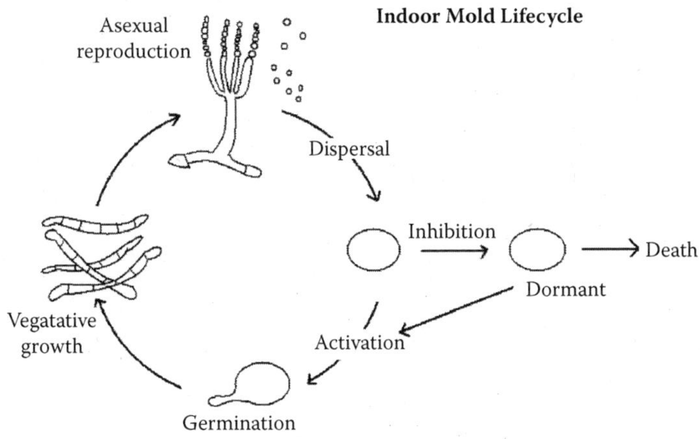

FIGURE 3.1 The basics of asexual reproduction.

3.3.3 Reproduction

As mentioned in Section 3.1, the asexual reproduction is the primary means of replication for indoor molds. However, it is important to note that a few indoor molds, such as *Eurotium amstelodami* and *Chaetomium globosum,* undergo sexual reproduction (Samson et al., 2002; Vesper et al., 2007a).

The asexual life cycle of indoor mold is illustrated in Figure 3.1. Let's begin the cycle, arbitrarily, with a mature spore-bearing structure. Any physical disturbance, such as vibration, flux in atmospheric pressure, shifts in relative humidity, or air movement, is enough to cause detachment of the spores from the spore-bearing structure. Once released, spores may be carried on air currents, transported by water, or carried by insects to new indoor surface substrates. Or spores may drop directly from the spore-bearing structure directly onto the previously colonized surface. Regardless of location, once a spore comes to rest, one of two phenomena may occur. The spores become dormant or they are activated. Spores remain dormant due to unsuitable substrate or environmental conditions. Nutrition, water activity, and competition all play roles in spore dormancy. However, if spores land on suitable substrate that is conducive to growth, they are activated and proceed to germinate. Germinated spores produce a hyphal network, which is like flexible tubes. The hyphal network becomes denser and forms the body or a vegetative mycelium of the mold. The mycelium produces enzymes that digest the indoor substrates in order to acquire nutrients for further growth. At some time during a mold's vegetative growth, an environmental trigger, often scarcity of nutrients or water availability, induces asexual reproduction. Specialized structures called conidiophores (ascomycetous molds) or sporangiophores (zygomycetous molds) arise from the vegetative mycelium and produce an abundance of spores. From here the cycle begins again, and may occur over and over many times, often producing billions of spores on water-damaged indoor substrates.

The basidiomycetes molds (typically yeasts) do not typically produce a vegetative mycelium, but instead reproduce via binary fission, where one mother cell splits into two daughter cells. The vegetative cells may grow in thin biofilms and usually require substrates that are water saturated or substrates that experience cyclic moisture or humidity saturation, such as showers and bathtubs.

Myxomycetous molds reproduce similar to ascomycetous molds, except individual spores germinate into ameboid cells that graze on bacteria-colonizing, water-saturated surfaces. Chemical pheromones released by the amoeboid cells signal the reproductive phase in the mold's life cycle. The ameboid cells congregate to form spore-bearing sporangiophores. Some amoeba become the stalk of the sporangiophore, while others become the spores.

3.3.4 Spore Dormancy and Germination

All mold spores experience a period of dormancy that occurs during the period of time from spore release to germination. Dormancy may last from a few hours to months or years. However, the longer the spore remains dormant the less likely it is to successfully germinate.

Dormancy is a survival mechanism that allows the fungus to remain in a state of suspended animation until its environment becomes hospitable. Some fungi have evolved extreme mechanisms that maximize success upon germination, such as dung fungi, which must pass through a gut of a ruminant before germination is possible. Indoor molds are generalists by nature and usually only require the correct combination of moisture, nutrients, and a suitable surface to activate and break dormancy. Many mold spores will germinate in nutrient-free water if placed on a surface to which it can adhere. Indoor mold species produce copious quantities of spores and are gambling that some spores will find nutrients upon germination in a moist substrate.

A dormant spore undergoes a series of complex biochemical, molecular, and structural changes before a germ tube is produced (Osherov and May, 2001). The spore coat glucans and proteins facilitate hydration and surface adhesion. Upon sufficient hydration, the molecular and biochemical machinery of the spore is initiated and controlled by the spore nucleus. Trehalose, a storage sugar in the spore, provides the necessary metabolic energy during spore activation. The nucleus directs mRNA to create new proteins that facilitate further hydration and ensure spore adhesion and capture of exogenous spore nutrients. Finally the spore reaches a point when the trehalose supply is nearly exhausted. This occurs in a matter of one to three hours depending on the species. Concurrently, the spore nucleus has ordered the building of a structural protein infrastructure and an enzymatic arsenal that delivers a feeding tube into the surrounding substrate. The germ tube carries the nucleus, which will replicate and perpetuate the genetic blueprint of the fungus and maximize nutrient extraction from the substrate.

3.4 NUTRITION AND METABOLISM

Two broad categories define nutritional modes of living things. They include the heterotrophs and the autotrophs. Autotrophs, which include green plants and some bacteria, use energy from the sun (photosynthesis) or the energy of chemical reactions to drive their internal cellular machinery and produce simple sugars. Heterotrophs on the other hand, lack the ability to produce food for themselves and have evolved a myriad of enzymes to break down and extract nutrients from dead organic matter

(Webster and Weber, 2007). Molds belong to the latter group of heterotrophs and secrete enzymes into the immediate environment that break down dead organic matter. They feed by reabsorbing the nutrients released during enzymatic degradation.

Molds extract nutrients from the surface substratum. They convert nutrients into fungal biomass and asexual spores. Molds use primary metabolism to secrete enzymes that are capable of breaking down cellulose or plant-derived materials (particle board, wall paper, drywall, tackboard) and petroleum-based polymeric materials (plastics, glues, chauking, carpeting, rubber, vinyl). The molds have a diverse suite of enzymes that allow them to degrade most indoor materials. Moreover, they are capable of growing on the dust particles that settle indoors. These particles are rich in cellulose, polymers, and mammalian cells.

Indoor environments are rich in easy-to-consume materials. Just like processed food that has been refined to maximize caloric power and reduce cost, most building materials are made from processed wood materials. Natural woods and plant materials have antifungal compounds that the plant produces to protect itself from a barrage of plant pathogenic fungi that constantly attack plants during its life cycle. Moreover, many recalcitrant structural compounds, such as lignins and phenolics, are difficult for molds to digest. However, in an effort to produce affordable building materials, most products are pulped, refined, and mixed into slurries that can be shaped into woodlike materials. During refining, the protective chemicals in the plant material are destroyed and the recalcitrant compounds are removed. In the end, a perfect growth medium for mold is produced.

3.5 TOXIN PRODUCTION

Toxin production in most indoor molds is a function of secondary metabolism. Unlike primary metabolism, which is always switched on in a growing mold, secondary metabolism is activated only when a mold is faced with certain environmental challenges. Water-soaked materials are an excellent environment for many other organisms to colonize in addition to mold. Some of these organisms, such as mites, feed on molds. They graze on molds like cows graze on grass and, like many types of plants that have evolved genes that induce the production of secondary metabolite antigrazing chemicals, molds, too, have evolved similar genes that are most commonly known as toxin genes. These toxin genes produce a variety of mycotoxin to deter grazing by insects and repel competitor molds.

3.6 SEQUENCING OF TAXONOMIC-RELEVANT DNA REGIONS

The nuclear-encoded ribosomal RNA genes (rDNA) of fungi have proven to be the most useful in the amplification and sequencing of most major groups of fungi. This rDNA region exists as a multiple-copy gene family comprised of highly similar DNA sequences. This similarity of DNA sequences gives way to the ability to use conserved primer sequences with the ability to identify fungi down to the species level and sometimes even differences within species.

The most common region of fungal DNA sequenced is perhaps the internal transcribed spacer region. It is most useful for molecular systematics at the species level primarily due to the higher degree of variability than the other genic regions of rDNA, for example, the large subunit (LSU) rDNA region. Most laboratories only use the first 600 to 900 bases of the LSU gene because this is the most variable region. Much of the LSU gene is invariant even across highly divergent taxa. Therefore, it is important to know how specific of a primer set is needed for the sequencing task at hand; the more variant the DNA region is, the more specific the primer set for that sequence will be.

A useful primer set used for fungal identification is the ITS1 (TCCGTAGGTGAACCTGCGG) and ITS4 (TCCTCCGCTTATTGATATGC) primer pair. As described earlier, this pair flanks the internal transcribed spacer region of high variability. Utilizing this primer set allows for the identification of fungi not only at the genus level but more often enough the species level.

3.6.1 Mold Identification Using the NIH GenBank Sequence Database

After an ideal primer set has been chosen and the DNA of interest amplified and sequenced, it is time to visually inspect the quality of the sequenced data, which is done by inspecting the chromatogram. Every nucleotide base from the sequence can be visualized by the different color-coded peaks. Adenine is green, thymine red, guanine black, and cytosine blue.

The major challenges with sequenced data is that the first 15 to 40 bases generally are of poor quality in addition to the fact that sequences tend to deteriorate after about 700 to 900 bases. This is why it is a good idea to sequence your DNA with a forward and a reverse primer in hopes of sequencing the entire polymerase chain reaction (PCR) product. There are alignment programs available that can also aid with putting the two sequences together to form one large sequence. It also needs to be noted that many sequencing facilities do not recommend using the same primer set used for amplification due to background noise of primer dimers. If you use a quality nucleotide clean-up kit after amplification to remove residual primers from your product, the general problem of background noise from your primers can be reduced and in most cases diminished.

Once the poor quality regions of the sequenced data have been removed, it is time to run a basic local alignment search tool (BLAST) to compare the sequence against other known sequences. The best database of sequences to date belongs to the NIH GenBank (http://blast.ncbi.nlm.nih.gov/Blast.cgi). Here the edited sequenced data can be input in hopes of finding a highly similar known sequence. For fungal sequences, a basic nucleotide BLAST is best. In most cases you can upload the edited sequence right into the query box from your desktop computer. If this poses a challenge, then manually enter the data paying attention to the FAST-ALL or FASTA format. A sequence in FASTA format begins with a single-line description, followed by lines of sequence data separated by a greater than (>) sign. If a line description is not necessary, the bare sequence can be pasted into the query box alone. Once the edited sequence is pasted, a job title can be chosen as well. The search set should be "others-nucleotide collection" since the mouse genome or human genome would not be of much use for fungal DNA. The program selection would be a "megablast-highly similar sequences." Now that the search parameters have been set, click "BLAST" and wait for the results to appear.

3.6.2 Genus-Level Identification

After the search results have been retrieved, that is, plausible matches for your sequence, genus identification should be pretty evident. Results with the greatest number of base pair matches in succession are your prime suspect for genus-level identification. For example, you would not choose a genus that out of a 600-base query search matched 100 percent for 70 bases as opposed to another genus that matched the query at 98 percent with 545 bases.

3.6.3 Species-Level Identification

Species-level identification may not be as simple as the genus-level identification. Closely related species tend to have highly similar sequences. Therefore, it may be necessary to research other differences between the species that can be tested within the laboratory setting, for example, microscopic examination and differential media.

3.6.4 Probes and Primers

The EPA has developed a broad offering of patented DNA probes and primer sets (www.epa.gov/nerlcwww/moldtech.htm). Patent 6,387,652 describes mold DNA sequences that make identification and quantification possible. MSQPCR technology has been licensed by United States and international firms. The patented technology includes probes and primer sets for over 130 of the major indoor air fungi (Table 3.1). MSQPCR technology strives to eliminate antiquated technologies, such as spore trap analysis, and to bring mold detection into the realm of medicine and biotechnology.

Each probe is a unique sequence of DNA ~10 base pairs long. The probe has a fluorescent molecule attached that is cleaved off during PCR. Each PCR cycle generates more fluorescent signal release from the probe. The brightness of the signal (measured in relative fluorescent units) is compared to a standard fluorescent curve of mold species spores to calculate the quantity of spores present of that species in the sample.

3.6.5 Reagents, Master Mix, and Disposables

MSQPCR is very similar to the classical PCR method. A template DNA is amplified using DNA polymerase, a forward and reverse primer, deoxynucleoside triphosphates (dNTPs), buffer, and magnesium. In addition to these PCR components, a fluorescent probe is used to visualize DNA amplification in a real-time setting. Many real-time assays are comprised of a master mix that already includes buffer, dNTPs, an ideal concentration of magnesium, and a fast-start polymerase. Depending on the master mix, the incubation time for the DNA polymerase to become activated varies so make sure to read the specifications carefully.

Many assays for MSQPCR can be made ahead of time and kept at 4°C for up to one month depending on the master mix used. The assays are composed of the probe, forward and reverse primer, master mix, and PCR grade water to bring the assay up to the desired reaction volume. The only thing left to add before placing the reaction into the real-time instrument is the template DNA.

All disposables should be DNA, RNA, and nuclease free. This will reduce the likelihood of contaminating DNA finding its way into the MSQPCR reaction as well as loss of template DNA from nucleases. Gloves should also be worn when dealing with DNA to keep the nucleases from degrading the template DNA.

3.6.6 Standards and Quality Control

In order for quantitative PCR to work, known concentrations of standards are needed. Standards for use in defining standard curves should be between 10^6 to 10^8 spores per milliliter. Using the known spore concentration, perform serial dilutions to a final dilution of 10^0, which should be between 1 and 9 spores per milliliter. Standard curves should be replicable with at least five points per dilution and reactions should mirror the reactions being processed daily. For example, the same master mix, run parameters, primer, and probe concentrations as well as manufacturer, and reaction volume should all be the same for both the standard and the everyday reaction. Any time a master mix or primer and probe manufacturer is changed, new standard curves are necessary. Small changes in the slope of the curves from these differences can drastically change the amount of mold calculated from the curve analysis.

It is also necessary to quality control every new lot of reagents used in the MSQPCR process. This includes the extraction kit, tubes, tips, master mix, and completed assays. This will ensure that no contamination is detected from the NIC (no inoculum control) and all the assays will detect mold as determined by the positive control for each species' assay. If by chance an amplification curve appears within the NIC, this value should be subtracted from the final calculation of spores or the assay can be remade; although remaking the assay is not the most cost-effective choice. If a positive control fails to yield an amplification curve, the assay can be rerun to ensure the primers and probes will not detect the mold present. If amplification fails a second time, the assay will need to be remade and another NIC and positive control will need to be run on this new assay.

3.6.7 Curve Analysis

Curve analysis requires two parts: visual inspection by the person running the sample and statistical analysis by the real-time machine software. Statistical analysis is done automatically by the machine and will yield a crossing-point (CP) or crossing-threshold (CT) value. This value will be used in calculating total spores present. Visual inspection is needed due to background fluorescence of the PCR plate. Highly sensitive software may pick up on background noise and call a curve positive that looks like nothing more than scribble. Without visual inspection, a curve, and subsequently a mold species, may be reported as present when indeed it is not.

3.6.8 Data Reporting and Limit of Detection

Once the curves have finished undergoing quality control, it is time to report the findings. As mentioned earlier the CP or CT values are used in calculating total spores present. Using the absolute quantification equation,

$$y = m (\log x) + b,$$

where

y = CP of the unknown sample,
x = concentration of the species' spores in the unknown sample, and
b = y-intercept of the species' calibration curve,

a spore concentration can be achieved with a simple manipulation of the equation

Spore concentration = 10 ^ [(CP – b)/m].

The limit of detection is based on the standard curves discussed previously. The minimum number of spores that are able to be detected based on the known concentration of spores detected in the serial dilutions of the standard is your limit of detection. For example, if a standard curve was used with an initial concentration of 1.59 × 10^6 spores per milliliter, the minimum number of spores detected at 10^0 would be 1.59 spores per milliliter.

3.6.9 Processing Other Sample Matrices

As previously stated, the ideal sample matrix for MSQPCR as set forth by Steven Vesper is house dust. Sampling using swabs or bulk material will require adjustments to the extraction process as well as the calculation of spores present. Swabs and bulk material may contain hyphal fragments that cannot be accounted for in standard curves. Hyphae can contain anywhere from one to hundreds of nuclei full of genomic DNA. Therefore it is not recommended to try to quantitate samples of this nature. Instead the MSQPCR process can be used strictly as a presence or absence diagnostic tool.

3.7 INTEGRATING GENE SEQUENCE DATA WITH MORPHOLOGICAL AND PHYSIOLOGICAL CHARACTERS TO OBTAIN HIGH-CONFIDENCE IDENTIFICATION

A database is only as good as the data it contains. GenBank is no exception. This is not a fault of the database managers, but fault of those who submit DNA sequences. Good sequence data is easy to get with the myriad of excellent DNA extraction kits on the market. However, ensuring that the DNA sequenced is from the target organism requires skills and knowledge of molecular biology beyond simple extraction kit, cookbook-like instructions. In reverse, a scientist may be a PhD molecular biologist, but know nothing of the morphology or biological characteristic of the organism he or she is working with. This is particular problematic for fungi. Since most molds grow on standard media, nonmicrobiologists often end up with contamination. If that contaminant mold is extracted and sequenced rather than the target mold or fungal culture, the error is propagated with the DNA into the sequence database. Even expert mycologists have difficulty distinguishing similar mold species that they have not worked with before. This is especially true for the *Aspergillus* and *Penicillium* genera. This means relying solely on GenBank or other sequence databases for genus and especially species identification is dicey at best. The preferred method is to integrate DNA with classic morphological and physiological identification methods. The hybrid of that marriage provides a powerful tool to identify unknown mold isolates.

3.8 MOLD SPECIFIC QUANTITATIVE POLYMERASE CHAIN REACTION (MSQPCR)

Low-cost screening techniques such as a visual walk-through and spore traps are often ineffective methods for detecting moderate or hidden molds issues. A recent study conducted by the EPA and the Department of Housing and Urban Development found that visual inspection alone failed to identify mold in homes that were known to have significant mold contamination due to water intrusion (Vesper et al., 2009). A paradigm shift from microscopic exam to DNA analysis and air samples to dust samples is required to provide adequate mold detection in the majority of American homes.

MSQPCR was developed by the EPA to identify and quantify abnormalities in the mold ecology of a home. A list of species identified by EPA-patented technology is provided in Table 3.2. The technique takes advantage of advance DNA detection technologies used in the pharmaceutical and biotech industry. MSQPCR is a method that uses species-specific DNA probes that fluoresce in the presence of the target mold. The more fluorescence the more spore of a particular species that are present. That is an oversimplification of the process, however. Fluorescence detection and probe cocktails take place in the laboratory. Liquid-handling robotics and advanced quantitative PCR platforms such as the Roche Lightcycler 480 handle most of the laboratory analysis.

3.8.1 Choice of Real-Time Instrumentation

There are more choices for real-time or quantitative PCR (QPCR) platforms available today than ever before. The important thing to consider when choosing a QPCR machine is throughput and reliability. The system must be able to analyze microtiter plates. Some systems still use glass capillaries and plastic tubes. These systems are often very reliable but they can only run 70 to 90 reactions at a time. Most MSQPCR analyses are panels of species, such as the Environmental Relative Moldiness Index (ERMI; Vesper et al., 2007a) that incorporates 36 mold species plus a control species, and there are several no inoculum controls (NICs) that must run with each sample also. Since each mold species takes up a spot in the QPCR machine, space becomes valuable. Even the smallest size microtiter plate that has 96 wells only allows for two environmental samples. The step up is a microtiter plate with 384 wells. This allows for running batches of 10 samples at a time. Moreover, microtiter plates are easily manipulated by liquid handling robots. Using a robot significantly increases

TABLE 3.2
List of Species Identified by U.S. EPA Patented MSQPCR Technology

Absidia corymbifera	Candida maltosa	Penicillium brevicompactum	Penicillium verrucosum, svar. 1
Acremonium strictum	Candida parapsilosis	Penicillium stoloniferum	Rhizomucor meihei
Alternaria alternata	Candida sojae	Penicillium canescens	Rhizomucor pusillus
Aspergillus auricomus	Candida tropicalis	Penicillium chrysogenum, svar. 2	Rhizomucor variabilis
Aspergillus caespitosus	Candida viswanathii	Penicillium chrysogenum	Rhizopus azygosporus
Aspergillus candidus	Candida zeylanoides	Penicillium griseofulvum	Rhizopus homothalicus
Aspergillus carbonarius	Chaetomium globosum	Penicillium glandicola	Rhizopus microsporus
Aspergillus cervinus	Cladosporium cladosporioides, svar. 1	Penicillium coprophilum	Rhizopus oligosporus
Aspergillus clavatus	Cladosporium cladosporioides, svar. 2	Penicillium expansum	Rhizopus oryzae
Aspergillus giganteus	Cladosporium herbarum	Penicillium citreonigrum	Rhizopus stolonifer
Aspergillus flavus	Cladosporium sphaerospermum	Penicillium citrinum	Scopulariopsis asperula
Aspergillus oryzae	Emericella nidulans	Penicillium sartoryi	Scopulariopsis brevicaulis
Aspergillus fumigatus	Emericella rugulosa	Penicillium westlingi	Scopulariopsis fusca
Aspergillus flavipes	Emericella quadrilineata	Penicillium coprophilum	Scopulariopsis brumptii
Aspergillus niger	Emericella variecolor	Penicillium corylophilum	Scopulariopsis chartarum
Aspergillus awamori	Eupenicillium crustaceum	Penicillium crustosum	Scopulariopsis sphaerospora
Aspergillus foetidus	Eupenicillium egyptiacum	Penicillium camembertii	Stachybotrys chartarum
Aspergillus phoenicis	Eurotium amstelodami	Penicillium commune	Trichoderma asperellum
Aspergillus niveus	Eurotium chevalieri	Penicillium echinulatum	Trichoderma atroviride
Aspergillus ochraceus	Eurotium herbariorum	Penicillium solitum	Trichoderma citrinoviride
Aspergillus ostianus	Eurotium rubrum	Penicillium decumbens	Trichoderma hamatum
Aspergillus paradoxus	Eurotium repens	Penicillium digitatum	Trichoderma harzianum

(continued)

TABLE 3.2 (CONTINUED)
List of Species Identified by U.S. EPA Patented MSQPCR Technology

Aspergillus parasiticus	Epicoccum nigrum	Penicillium expansum	Trichoderma koningii
Aspergillus sojae	Fusarium solani	Penicillium fellutanum	Trichoderma longibrachiatum
Aspergillus penicillioides	Geotrichum candidum	Penicillium charlesii	Trichoderma viride
Aspergillus puniceus	Geotrichum klebahnii	Penicillium glandicola	Ulocladium atrum
Aspergillus restrictus	Memnoniella echinata	Penicillium griseofulvum	Ulocladium botrytis
Aspergillus caesillus	Mucor amphibiorum	Penicillium implicatum	Ulocladium chartarum
Aspergillus conicus	Mucor circinelloides	Penicillium islandicum	Wallemia sebi
Aspergillus sclerotiorum	Mucor hiemalis	Penicillium italicum	
Aspergillus sydowii	Mucor indicus	Penicillium melinii	
Aspergillus tamarii	Mucor mucedo	Penicillium miczynskii	
Aspergillus terreus	Mucor racemosus	Penicillium olsonii	
Aspergillus unguis	Mucor ramosissimus	Penicillium oxalicum	
Aspergillus ustus	Myrothecium verrucaria	Penicillium purpurogenum	
Aspergillus versicolor	Neosartorya fischeri	Penicillium raistrickii	
Aspergillus wentii	Paecilomyces lilacinus	Penicillium restrictum	
Aureobasidium pullulans	Paecilomyces variotii	Penicillium roquefortii	
Candida albicans	Penicillium aethiopicum	Penicillium sclerotiorum	
Candida dubliniensis	Penicillium atramentosum	Penicillium simplicissimum	
Candida glabrata	Penicillium aurantiogriseum	Penicillium ochrochloron	
Candida guilliermondii	Penicillium freii	Penicillium glabrum	
Candida haemulonii, svar. 1	Penicillium hirsutum	Penicillium lividum	
Candida haemulonii, svar. 2	Penicillium polonicum	Penicillium pupurescens	
Candida krusei	Penicillium tricolor	Penicillium spinulosum	
Candida lipolytica	Penicillium viridicatum	Penicillium thomii	
Candida lusitaniae	Penicillium verrucosum, svar. 2	Penicillium variabile	

accuracy and throughput by removing human error and saving considerable loading time.

The Roche Diagnostics Lightcycler 480 (LC480) system is an extremely robust system that incorporates both a 96 well microtiter plate format and a 384 well format. Moreover, they have recently released a new block for the LC480 that will analyze 1,536 well microtiter plates. That will allow labs to decrease reagent volumes (which decreases cost per sample) and run 30 samples at a time. The LC480 is packed full of software features that significantly reduce the burden of DNA analysts that must review and analyze the data output of each sample.

Another system used for MSQPCR is Applied Biosciences' 7900HT system. It will analyze 384 well microtiter plates, but it makes use of a propriety robotics system and puts the platform over $100,000. The Bio-Rad CFX384 shows much promise for MSQPCR. It has a unique heating system that allows the user to create gradients and optimize probe and primer reactions quantitatively. Most labs have a separate PCR system, like a Stratagene Robocyler to optimize probe and primer mixes. Now it can all be done on a single Bio-Rad platform in real-time.

The Roche LC480 system and Bio-Rad CFX384 system can be used with a variety of liquid handling robots, which provides greater flexibility for laboratories. Eppendorf makes excellent liquid-handling robots such as the *epi*Motion 5075. These systems are fast and extremely reliable and capable of producing consistent coefficient of variation (CV) values under 1 percent for volumes less than 10 μl. Tecan and Hamilton also manufacture a wide variety of liquid handling robots that work with many QPCR platforms.

3.8.2 Mold Spore Standards

MSQPCR requires a standard curve to determine the cell equivalent of mold spore in an environmental sample. For each mold species included in an analysis, a calibration curve must have been generated to report spore concentration. The calibration curve is generated from real mold spores that the laboratory must harvest from cultures. Moreover, since the concentration reported is particular to spores, which is a fairly consistent unit from a genetic standpoint, the majority of fungal hyphae and mycelium must be removed. The U.S. EPA specifies that mold spore stocks be 95 percent free of vegetative material (Haugland et al., 2004). Hence, a series of filtration steps and cleanup methods are required to ensure meeting of EPA specifications. Laboratories that conduct MSQPCR analysis will have spore stocks preserved in ultracold freezers or liquid nitrogen for each probe they offer.

3.8.3 Commercially Available MSQPCR Kits

To date, only Roche Diagnostics has developed a MSQPCR kit. The kit contains the probes and primers for the 36 species of mold found in the U.S. EPA's Environmental Relative Moldiness Index (ERMI) panel and American Relative Moldiness Index (ARMI) panel. The kits are available in both a liquid and freeze-dried format.

REFERENCES

Garraway, M. O., and Evans, R. C. 1984. *Fungal Nutrition and Physiology*. New York: John Wiley & Sons.

Griffin, D. H. 1994. *Fungal Physiology*, 2nd ed. New York: Wiley-Liss, Inc.

Haugland, R. A., Varma, M., Wymer, L. J., and Vesper, S. J. 2004. Quantitative PCR analysis of selected Aspergillus, Penicillium and Paecilomyces species. *Systematic and Applied Microbiology* 27: 198–210.

Hawksworth, D. L., Kirk, P. M., Sutton, B. C., and Pegler, D. N. 1995. *Dictonary of the Fungi*, 8th ed. Surrey, UK: CAB International.

Macher, J., ed. 1999. *Bioaerosols: Assessment and Control.* Cincinnati, OH: ACGIH.

Mueller, G., and Schmit, J. 2007. Fungal biodiversity: What do we know? What can we predict? *Biodiversity and Conservation* 16: 1–5.

Osherov, N., and May, G. S. 2001. The molecular mechanisms of conidial germination. *FEMS Microbiology Letters* 199: 153–160.

Rhodes, J. C., and Brakhage, A. A. 2006. Molecular determinants of virulence in Aspergillus fumigatus. In *Molecular Principles of Fungal Pathogenesis*, J. Heitman, S. G. Filler, John E. Edwards, Jr., and A. P. Mitchell, eds. Washington, DC: ASM, 333–345.

Samson, R. A., Hoekstra, E. S., Frisvad, J. C., and Filtenborg, O. 2002. *Introduction to Food and Airborne Fungi*, 2nd ed. Ad Utrecht, Netherlands: Centraalbureau voor Schimmelcultures.

Vesper, S., McKinstry, C., Cox, D., and Dewalt, G. 2009. Correlation between ERMI values and other moisture and mold assessments of homes in the American Healthy Homes Survey. *Journal of Urban Health* 86: 850–860.

Vesper, S., McKinstry, C., Haugland, R., Wymer, L., Bradham, K., Ashley, P., Cox, D., Dewalt, G., and Friedman, W. 2007a. Development of an environmental relative moldiness index for US homes. *Journal of Occupational and Environmental Medicine* 49: 829–833.

Vesper, S. J., McKinstry, C., Haugland, R. A., Iossifova, Y., Lemasters, G., Levin, L., Hershey, G. K. K., et al. 2007b. Relative moldiness index as predictor of childhood respiratory illness. *Journal of Exposure Science & Environmental Epidemiology* 17: 88–94.

Ward, M. D. W., Chung, Y., Copeland, L. B., Selgrade, M. K., and Vesper, S. 2007. Indoor molds and allergic potential. *Journal of Allergy and Clinical Immunology* 119: S99–S99.

Webster, J., and Weber, R. 2007. *Introduction to Fungi,* 3rd ed. Cambridge, UK: Cambridge University Press.

Zorman, T., and Jersek, B. 2008. Assessment of bioacrosol concentrations in different indoor environments. *Indoor and Built Environment* 17: 155–163.

4 Products of Mold Associated with Sick Building Syndrome

Walter E. Goldstein, PhD, PE

CONTENTS

4.1 SUMMARY

This chapter is a short summary of the mold products being covered in all chapters of this book. Molds in the vegetative state (growing and propagating) form products. The reason for this may be preservation and as defense measures. Besides the mold particles themselves, these products can be very hazardous to health and damaging to building structures.

Table 4.1 presents a sample of mold types and products that have been associated with sick building syndrome. The products that can be damaging to health include mold fragments, which may lodge in airways, cause inflammation, or immune system responses (Croft et al., 1986; Flanigan, 1987; Ghorai, 2009; Kercsmar et al., 2006; Kim et al., 2010; Nielsen, 2004; Jang, W-K et al., 2008). They also include secondary metabolites (products of the mold) that one party estimates to be 300 in number (Gutarowska and Piotrowska, 2007). In sufficient quantity, they can be fatal or at the least cause serious illness. Gutarowska and Protrowska, note that mold spores <10μm can reach far into respiratory passages. There are organic molecules that can irritate tissue (oral, lung, skin) and cause reactions (Hajdu, 2009). Dome studies relate many species of mold to secondary metabolites of various kinds, pointing out potential dangers to health (Fischer et al. 2000; Shoemaker and House, 2005).

Typically, the mold types are reported, the secondary metabolites are recorded, and other entities may be reported. Other entities may be fragments of mold particles that are allergenic and also breakdown fragments of wallboard such as beta D glucans, long chain carbohydrates that may be formed by the mold breaking down the cellulose, or else in the cell wall of the mold itself.

There has not been a systematic study relating the mold organism, its growth, its reaching a stage to form metabolites, and the characterization of such metabolites.

TABLE 4.1

Organism, Metabolite, and Disease Association

Organism Genus	Species	Secondary Metabolite	Disease	Other	Treatments in Literature	Reference
Cladosporium	cladosporiodides					
Alternaria	alternata		red cell hemolysis due to hemolysin			Vesper and Vesper (2004)
					posaconazole, voriconazole	Hajdu et al. (2009)
Penicillium	chrysogenum		mucous membrane			Nielsen (2004)
			red cell hemolysis due to chrysolysin			Vesper and Vesper (2004)
	oxicalis					
	mereagrinum		leukemia association			
Stachybotrys	chartarum		mucous membrane			
			asthma, airway inflammation, rhinitis			
	atra	trichothecenes	upper respiratory	hay, straw		Flanigan (1987)
					CSM therapy (cholestryramine)	Shoemaker and House (2005)
						Croft et al. (1986)
				water damaged		Nielsen (2004)

			red cell hemolysis due to stacylysin			Vesper and Vesper (2004)
Aspergillus	flavus	aflatoxin		corn, peanuts	itraconazole, echinocandins	Hajdu et al. (2009)
	fumigatus			wood dust		
	parasitius	aflatoxin zearalonone	leukemia immosuppressiveness			
	versicolor		cancer mycotoxin sterigmatocystin			Nielsen (2004)
			asthma, airway inflammation, rhinitis			Kim et al. (2010)
Trichderma	spp.		mucous membrane			Nielsen (2004)
			T-2 toxin?	wood		
Chaetomium	spp.		mucous membrane			Nielsen (2004)
all		beta d glucans	respiratory inflammation from fungi or breakdown of cellulose			
all					immune stimulants such as Tuber melanosporum (truffles)	Ghorai et al. (2009)

This needs to be done in a way leading to a set of diagnostics that can tie organisms, metabolites, and maybe conditions together so that possibly a therapy can be suggested and tried. The work in this book is a beginning for that. The references all have useful information. However, unless they are tightly linked, it is difficult to progress and learn what may be done about damaged property, ways to prevent damaged property, ways to detect health-causing problems, and how to more precisely treat health concerns.

Clearly, new work is needed to understand how molds biomanufacture products, such as trichothecenes, in the hope such work can lead to ways to prevent such product formation and in addition inhibit vegetative growth from sporesor limit sporulations. Research at the U.S. Department of Agriculture (USDA) reports developing strains of the fungus, Myrothecium verrucaria, to infect a plant pest, the flowering ornamental kudzu (Boyette, 2009). Release of the fungus is intended to destroy the invading kudzu plant. The fungus, unfortunately, produces trichothecene compounds (toxins). The USDA workers noted the necessity to reduce or stop the production of trichothecenes. The USDA workers reported using a proprietary distilled water method called "spore washing" to prevent the mold from producing the trichothecenes. They also found that when using liquid rather than solid fermentation culture, they silenced the trichothecene production or reduced the production significantly. There has also been work on developing trichothecene-resistant transgenic plants (Hohn, 2006). As noted in this book, trichothecenes are a dangerous class of products also produced by mold infecting wallboard. These steps may in some fashion be applicable to wallboard situations to prevent mold and the formation of mold products.

Sometimes, a negative factor such as trichothecenes can become beneficial. In this case, for example, one activity of trichothecenes is mentioned in specifications in a patent noting use of such compounds to prevent reformation of tissue-blocking arteries following angioplasty and perhaps stent insertion (Kunz et al., 1999). The compounds reported inhibit smooth muscle formation. Inhibiting smooth muscle formation unintentionally may be a health-negative consequence of human exposure to trichothecene. However, the activity may be representative of a way to improve cardiovascular surgery. Therefore, research in the area to block trichothecene production may result in findings that transfer to the other areas for commercial benefit. The example (in this case to the medical area) is notable for enterprising companies since such events often result from research.

Perhaps these methods used at the USDA inherently involve an inhibitory step or steps that can prevent mold production of trichothecene in wallboard to at least alleviate part of the problem. Perhaps the mold growth or its sporulation or its generation into vegetative form can be prevented through related inhibitory steps, which may be another product of mold fermentation.

REFERENCES

Boyette, C. 2009. Formidable fungus goes toe to toe with kudzu. *Agricultural Research* 57(9) (July): 4–5.

Croft, W., Jarvis, B. B., and Yatawara, C. S. 1986. Airborne outbreak of trichothecene toxicosis. *Atmospheric Environment* 20(3): 549–552.

Fischer, G., Muller, T., Schwalbe, R., Ostrowski, R., and Dott, W. 2000. Species-specific profiles of mycotoxins produced in cultures and associated with conidia of airborne fungi derived from biowaste. *International Journal of Hygiene and Environmental Health* 203: 105–116.

Flanigan, B. 1987. Mycotoxins in the air. *International Biodeterioration* 23: 73–78.

Ghorai, S., Banik, S. P., Verma, D., Chowdhury, S., Mukherjee, S., and Khowala, S. 2009. Fungal biotechnology in food and feed processing. *Food Research International* 42: 577–587.

Gutarowska, B., and Piotrowska, M. 2007. Methods of mycological analysis in buildings. *Building and Environment* 42: 1843–1850.

Hajdu, S., Obradovic, A., Presterl, E., and Vécsei, V. 2009. Invasive mycoses following trauma. *Injury* 40: 548–554.

Hohn, T., Peters, C., and Saleron, J. May 23, 2006. Trichothecene-resistant transgenic plants. Assignee: Syngenta Participations, AG, U.S. Patent 7,049,421.

Jung, W. K., Lee, D. Y., Choi, Y. H., Yea, S. S., Choi, I., Park, S. G., Seo, S. K., et al. 2008. Caffeic acid phenethyl ester attenuates allergic airway inflammation and hyperresponsiveness in murine model of ovalbumin-induced asthma. *Life Sci.* 26; 82(13-14):797–805.

Kercsmar, C., Dearborn, D. G., Schluchter, M., Xue, L., Kirchner, H. L., Sobolewski, J., Greenberg, S. J., Vesper, S. J., and Allan, T. 2006. Reduction in asthma morbidity in children as a result of home remediation aimed at moisture sources. *Environmental Health Perspectives* 114(8):1574–1580.

Kim, C. W., Seo, J. S., Kim, M.-K., Jun, J. E., Choi, J. C., and Choi, B. W. 2010. Secondary cutaneous aspergillosis disseminated from the lungs of a patient with asthma on 1 month steroid treatment. *Diagnostic Microbiology and Infectious Disease* 66: 104–107.

Kunz, L., Klein, R., and Reno, J. 1999. Therapeutic inhibitor of vascular smooth muscle cells. U.S. Patent 5,981,568, November 9.

Lee, C-M, K. M, S-K, Jeon, Y-J, Choi, I-W. 2008. Caffeic acid phenethyl ester attenuates allergic airway inflammation and hyperresponsiveness in murine model of ovalbumin-induced asthma.

Nielsen, K. 2004. Mould growth on buildings under low water activities: Influence of humidity and temperature on fungal growth and secondary metabolism. *International Biodeterioration & Biodegradation* 54: 325–336.

Shoemaker, R. C., and House, D. E. 2005. A time-series study of sick building syndrome: Chronic biotoxin-associated illness from exposure to water-damaged buildings. *Neurotoxicology and Teratology* 27: 29–46.

Vesper, S. J., and Vesper, M. J. 2004. Possible role of fungal hemolysins in sick building syndrome. *Advances in Applied Microbiology* 55: 191–213.

5 Mathematical Model of Mold Propagation and Product Formation in Building Materials, Inherent Transport Phenomena, and Applications

Walter E. Goldstein, PhD, PE
Willem A. Schreuder, PhD

CONTENTS

5.1 DESCRIPTIVE MODEL OF MOLD CONTAMINATION

In this chapter, a mathematical model is developed that can be used to derive solutions to problems in mold contamination (based on inference and using experimental data derived from observations and associated analytical methods). The solutions will help remediation be more efficient and reduce costs, for example, by leading to suggestion of

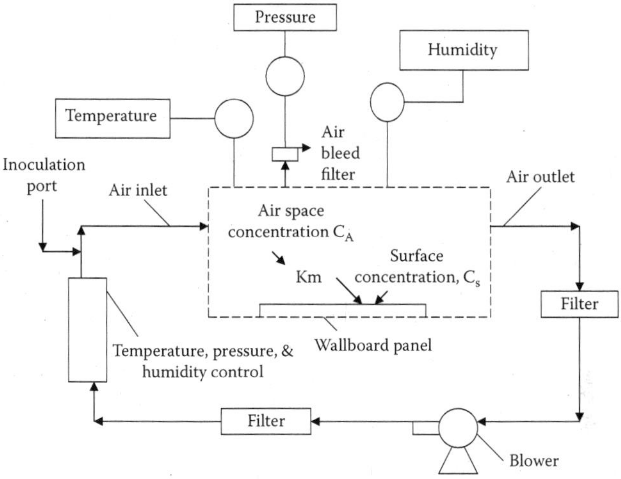

FIGURE 5.1 Air test chamber schematic.

construction materials with desired properties that reduce the magnitude of the problem. The property owner will benefit from use of the model by themselves or through professionals to help understand, correct, and prevent problems to make the best of an often difficult and costly situation. The methods taught by use of the model will be invaluable to insurance companies to reduce their risk and liability. The methods will allow definition of parameters and better understanding to treat health issues due to mold particles and their chemical products (both potentially harmful to health). The methods and equations can be used in a predictive manner—in effect, producing a valuable expert system.

Consider a space in a dwelling that has become (or is becoming) inundated with water intrusion, perhaps gradually (depicted as part of the experimental assembly to study the matter shown in Figure 5.1). Similar experimental assemblies have been described in the literature (Buttner et al., 1999, 2002). Further, suppose that the means to ventilate or dry the space is inadequate. The air that enters the space is laden to a given quantitative degree with dust particles and mold particles—the mold particles perhaps intermingled with the dust. A further complication might be paint flakes containing lead, which may commingle or even interact with the mold particles and mold itself (if, for example, the mold sequesters the lead in the process of degrading the paint and absorbing nutrients it requires for sustenance).

The mold-laden dust or mold particles present at some varying time-dependent concentration enter the area with some velocity distribution that can be quantified. Sampling devices (as covered in Chapters 6 to 9) can be used to collect samples by location and as a basis to determine the time dependence of particle size distributions (Ashley et al., 2008).

Samples can be analyzed though (1) microscopy for particle size distribution; (2) mycological classification; and (3) typed phenotypically and genetically so a complete picture results (Cruz et al., 2001; also see Chapter 8). Quantitative real-time polymerase chain reaction or DNA sequencing for species identification, and secondary metabolite analysis can be conducted as noted in Chapters 3 and 9 (Cruz et al., 2001; Gutarowska and Piotrowska, 2007; Haugland et al., 2002; Wady et al., 2003).

5.2 DISPOSITION OF MOLD-CONTAINING PARTICLES FOLLOWING ENTRY

Depending on velocity, turbulence (or lack of), quiescence, and particle size and concentration, the dust particles, mold spores, and paint flakes may alight on a vertical or horizontal surface such as wallboard (drywall). They will adhere to different degrees. There may be a charge association between mold particle and wallboard surface. Since the wallboard surface is typically composed of cellulose, it is nutritious for the mold. The mold spores then transform into vegetative growth, which can then propagate through the wallboard surface and perhaps penetrate into the gypsum sublayer.

5.3 MOLD RENEWAL

The vegetative mold, if subjected to drying and stress, may sporulate to protect itself as a seedlike entity. Such spores may spread anew and the process continued again (as spores are activated to enter vegetative growth again). The model encompasses and treats spore processes.

5.4 MOLD SECONDARY-PRODUCT HAZARDS

The mold produces secondary metabolites of a variety, which may be volatile organic compounds (VOCs) and variant biochemical/chemical molecules such as mycotoxins. Some may reside in the wallboard paper surface or its gypsum sublayer. Other molecules may transport to the air space.

The dangers of the molecules lie in ingestion by mouth causing gastrointestinal and other problems if passed internally from the gastrointestinal tract to surrounding tissue and the circulatory system. The dangers also lie in mold entering nasally, thereby causing pulmonary problems if captured by lung tissue, or else entering the body by skin intrusions. The nature of the array of mold products and the concentrations and particle size present at the surface and delivered to air space directly impact on inherent health hazards and thereby create a sick building (Institute of Medicine, 2004, 129). The model is used to predict such concentrations and thereby provide input for health care needs.

5.5 USE OF ANALYTICAL DATA FOR MODEL DEVELOPMENT

The identification of these particles and molecules by analytical means and measurement of their concentration results in markers for prognosis for the effects of the damaging agents and assessment of therapeutic avenues. These particles and molecules are also markers governing the potential to physically treat the contamination by nullifying the effects of the particle or chemical, breaking the particle or chemical down, or else simply reaching a threshold where the only solution possible in remediation is removal and evacuation of the building.

In order to apply the mathematical model, actual model surfaces are treated and studied in the laboratory and also examined in a predictive manner using the mathematics presented in this chapter. Such surfaces are purposely inoculated and studied in a controlled environmental chamber (Buttner et al., 1999). The experiments conducted in the laboratory mimic field circumstances. Field samples are compared to laboratory samples by observation, by microscopy, and by chemical and genetic means. The presence of mold products is measured in the laboratory and in the field.

The mold particles and entities produced by the mold particles are somewhat quantified in the literature. Values have been generated to test the mathematical model as reported in this chapter. Some of the products are mold volatile organic compounds (MVOCs), and some are classified as mycotoxins. Measurement of MVOCs may be typically done by gas chromatography, gas chromatography-mass spectrometry (GCMS), and possibly by electronic nose devices as available (Doleman et al., 1998; Wady et al., 2003).

Interestingly, industries with a vested interest in odor measurement may be able to offer synergistic knowledge and measurement technology. For example, odor measurement may be conducted to resolve a pleasant experience in differentiating wines. However, if one can detect something, it may or may not pose an alarm, and it may be harmful if exposure continues with the concomitant incapacity to detect the potentially harmful entity any longer (Delahunty et al., 2006). Since the model partly deals with detecting and quantifying mold particles (and identifying molds), it is very necessary to obtain quantitative data on products produced by the molds where possible (metabolites and mycotoxins) and compare the results to the literature since this data relates to health effects and structural damage.

Some studies present vapor pressure as a key measure of a detectable volatile (Covarrubias-Cervantes et al., 2004). This is useful baseline data. However, it is recognized that humans can detect the presence of an entity well below a reported saturated vapor pressure. In order to be detected, the entity must be volatile or otherwise delivered to air space. However, a vapor pressure measurement represents an equilibrium process between volatiles condensing or evaporating. Therefore, molecules are continually evaporating.

Although it is anticipated that humans can detect concentrations well below saturated vapor pressure values, it is not clear at what levels (either acute or chronic) that compounds such as those produced by molds pose health dangers. The data presented in this chapter and its representation by the model can be used by health care workers to assess ill effects in clinical studies. The specific classification of molds and association with the products they form is exceedingly important since, for example, MVOCs may number in the hundreds and many are reportedly produced by molds (Wady et al., 2003). The data and the mathematical model will also help assess the extent and nature of mold product concentrations and when to apply a specific remedy for a contaminated building.

5.6 ROLE OF SIMULATION (IN SILICO EXPERIMENTS WITH THE MODEL)

The model is developed by in silico (using the model and computer software) experiments as well as use of experimental data from the literature (Danchin et al., 1991). Through in silico experiments, the model can be tested for feasibility using appropriate inputted numbers to show it is internally consistent in predictions. The combination of simulation and treatment of literature data is used to establish the validity of the model. This approach aids in quality control of the research since simulation data and experimental data can be used as a means to cross-check consistency and accuracy. Instituting this program is an inherent quality assurance necessity when the model is applied.

The development of an expert system based on validation of this model will be important to health care needs and the economic well-being of the nation based on the addition of valued scientific and engineering information. Certainly, the authors of this book are a composite of industrial and academic experts that can input data to the model and also benefit from it. This work helps implement a policy to improve our nation's communities, including disadvantaged segments, since the model and data can be used to improve housing and the well-being of our citizenry.

5.7 DEVELOPMENT AND ACQUISITION OF DATA FOR THE MATHEMATICAL MODEL

5.7.1 Particle Influx and Capture

Particles enter an enclosure in an airstream of varying humidity, temperature, composition, and particle size distribution. The enclosure may be a laboratory unit, pilot plant unit, or a field installation.

As indicated in the sketch provided in Figure 5.1, depicting such an enclosure, particles laden in the gas stream enter the chamber and pass through it, with the opportunity to transfer to chamber walls outfitted on any surface with wallboard sections or pieces. The air inlet flow at a particular concentration of particles is assumed to be well mixed in the chamber. Perfect mixing is assumed in all directions so that gradients do not exist in the gas stream. If mixing is not perfect, then particle concentration will be dependent on chamber geometry—this is to be avoided to simplify interpretation of data obtained.

As shown in Figure 5.1 and Figure 5.2, the mass flow balance consists of the mass flow of particles entering equal to the mass flow of particles exiting, particles accumulating in unsteady state in the volumetric space (which may be a relatively small value), and the mass flow of particles transferring to the surface based on a mass transfer coefficient, K_m, encompassing convective and diffusive mass transfer (Bird et al., 2007). As the particles transfer to the wall, they will be depleted from the well-mixed airstream. The mold particles may carry nutrients and triggers for interconversion of spores to vegetative growth and also the reverse.

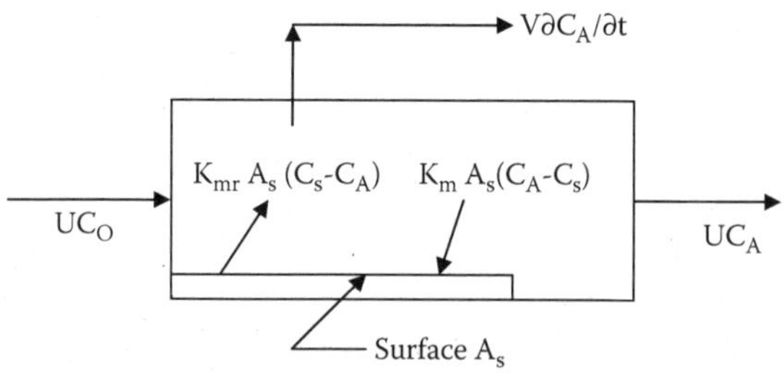

FIGURE 5.2 Mass balance for transport of particles in airstream to wallboard surface.

The changes in such entities may occur within any mold particle so that the distribution of spores and vegetative cells within the air space can vary as a function of time and conditions. It is expected, however, that such reaction and interconversion will occur primarily at the wallboard surface where nutrients are primarily located so such intraparticle reaction is not assumed.

5.7.2 Mold Particle Capture at the Wallboard Surface

At the surface, the particles (e.g., collection of spores forming a colony) will accumulate and may remain where they landed or migrate. In special cases, such as to the panel at the base of the chamber or where the air is especially quiescent, particles may transfer by diffusion and also at a velocity rate close to Stokes' law settling. If the particles are captured and bound to the surface, thereby preventing their release into gas space, then the effective concentration for transport of particles to the surface (at the surface) is zero. The particles will, however, collect at the surface in a manner that can be quantified. If the surface capacity is inadequate to accumulate the particles, then the transfer will be limited.

Given this description as a basis, there is a first partial differential equation depicting a mass balance over spores in the airstream element, accumulating in the airstream, and then losing particles from the airstream to the surface (terms as shown in Figure 5.2). The partial differential equation (rather than ordinary differential equation) is necessary as the variables such as particle concentration are a function of position as well as time. The process is unsteady state (time variable) in nature. The limit of this partial differential could end up being a constant that is approximately equal to the value of the particle concentration at the entry point to the chamber (assuming the loss of particles to the surface is relatively small). A partial differential equation exists for the spores and a second one for the vegetative cells.

5.7.3 Equations for Particles and Spores at the Wallboard Surface

A third partial differential equation covers the time-dependent accumulation of particles at the surface resulting in part from the net addition of particles to the surface from the airstream governed by the mass transfer coefficient (K_m), the depletion of the particles through spore migration, the depletion due to potential conversion of the spore particles to vegetative cells, and an increase in particle concentration due to the possible conversion of these vegetative cells back to spores. The basis for the third partial differential equation is depicted in Figure 5.3 as a shell balance of the mass transfer processes over a differentially thin circulator shell segment, Δr, that is also a differentially thin section, Δz, of a very thin cylindrical piece. The thickness of the thin cylindrical piece is the thickness of the wallboard surfaces plus substructure. The total thickness consists of paper surface of thickness, p; the sublayer gypsum plaster of thickness, l; and followed by paper surface thickness, p. The top and bottom surfaces may not be exposed to equivalent conditions that may cause mold. For example, the entire wallboard may be wet but airflow on one side may exceed airflow on the other side and the wallboard may therefore dry differently (if allowed to dry at all).

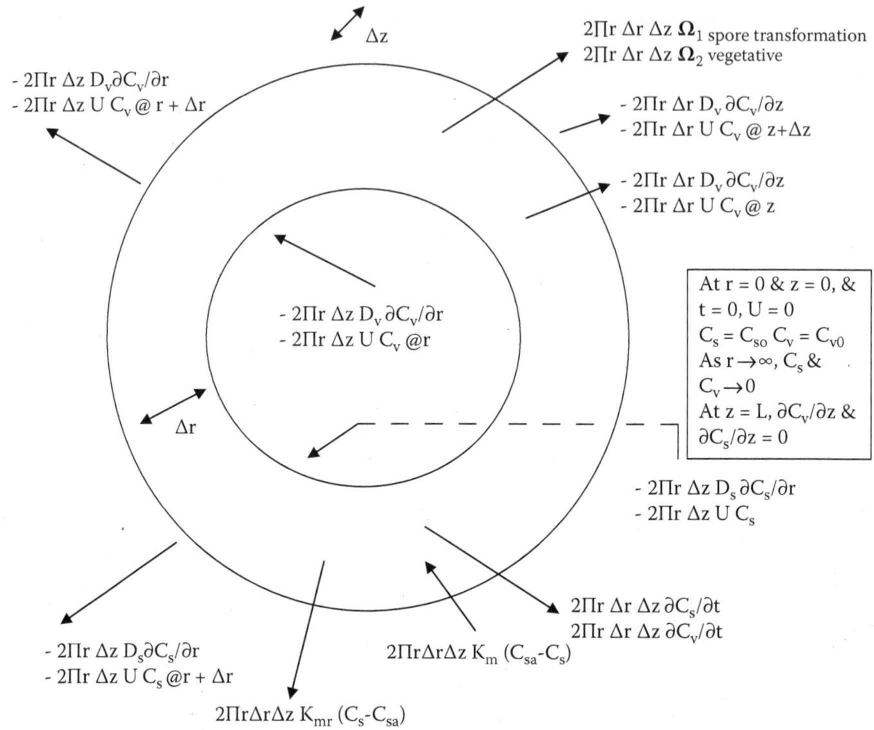

FIGURE 5.3 Shell mass balance over spore and vegetative growth complex space.

The processes proceed in a radial direction across the surface of the wallboard and also in an axial direction into the wall board. The processes occur for both spores and vegetative cells, requiring separate mass balances to describe them. The processes may also occur in the sublayer below the wallboard paper layers and may also be affected by air supply to the sublayer of the wallboard.

Mathematical treatments here follow practices in chemical engineering science for analysis of transport phenomena (Bird et al., 2007). Carrying out methods of calculus on this set of equations leads to partial differential equations. These partial differential equations are converted to numerical analysis forms using algorithms described further in this chapter.

The basis of a fourth partial differential equation is also shown in Figure 5.3 as the mass balance for the appearance of vegetative cells from spores and their possible disappearance back to spores, their accumulation, and their transport through the surface. Since vegetative cells may revert back to spores, the sources of the spores are this reversion as well as the addition of spores (from points of deposit on the surface and originally from the atmosphere). Some spores may of course transfer back to the atmosphere as indicated by the equation containing K_{mr}, the mass transfer coefficient for release of spores from the surface back to air space.

The transport is allowed to be both diffusive (D terms) and convective or velocity (U terms) to generalize the situation for a number of circumstances. The diffusive terms are also reflective of dispersion that may occur due to backmixing. Analysis of time-dependent migration data obtained from small- and large-scale laboratory chamber experiments in observing spore-vegetative processes at surfaces will allow one to judge if diffusive, dispersive, or convective effects are apparent. The C terms are either spore concentration (C_s) or vegetative concentration (C_v). The Ω terms represent biochemical reactions for loss or gain of spores or vegetative cells. The process starts at time zero and proceeds.

The conditions at time zero reflect the depiction of real-world situations that occur in mold contamination. For example, the area may initially be quite dry with mold particles, perhaps a mix of spores and vegetative cells present in the air space, Moisture is introduced into the air space through airflow, or through a pipe or valve suddenly starts leaking unbeknown to building occupants. This water intrusion may begin to wet and eventually saturate the wallboard. Mold then begins to propagate and cause damage. If the problem is discovered quickly enough, then drying can start.

The model can be used to evaluate the techniques including drying used in mold remediation. The timing on discovery of the problem and correction can be treated using the model. The model can be used to predict the level of mold contamination that is reached and the degree of remediation required and the potential danger to health due to mold particles and secondary metabolite formation. The model can be used to judge treatment options (for example, in situ eradication of mold perhaps lessening construction repair).

Reaction mechanisms and transport may be varied many ways to conduct simulations. For example, convection in the wallboard may be limited and diffusive processes may predominate. The gypsum sublayer may be so void of nutrients that although spores penetrate into the sublayer, they do not readily convert to vegetative cells there, or if they do, they revert back or die. Therefore, the sublayer may simply be a wet but inert layer. This can of course alter the rate of mold propagation and the rate of production of mold secondary metabolites.

5.7.4 Multiple Locales for Spore Contamination of Wallboard

Many spores land and deposit so there are multiple centers of radial segments. A center is referred to herein as spore and vegetative growth space. At some point, the outer radius is untouched by a growing circular segment, and that represents a zero time concentration boundary condition.

Figure 5.4 presents a sketch of a cross-section of wallboard (drywall) showing the top paper layer wrapped around a typically gypsum plaster inner core. The piece may be viewed to be symmetrical though conditions that are imposed may not result in all parts of the wallboard being exposed in the same manner. From the top surface (which can also be the first surface sideways), the paper layer is of thickness, p, and proceeds from z position zero to z_p. Since the sublayer (gypsum) is of thickness 1, half of this sublayer represents the point where half of the subsection ends at position $z_{p+1/2}$. The remaining half of the sublayer ends at z_{p+1}. The bottom (or backside) paper layer ends at z_{p+21}.

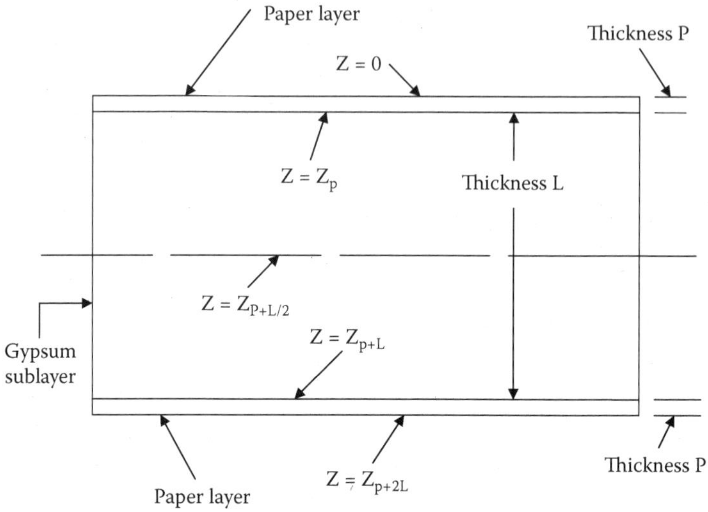

FIGURE 5.4 Wallboard cross-section.

The spores and vegetative growth can also penetrate into the board and substructure (the z direction). The z direction is bounded first by the end of the first paper layer at z_p and by halfway through the wallboard below the half substructure at $z_1$where transport may be zero ($\partial C_s/\partial z$ or $\partial C_v/\partial z = 0$). This represents symmetrical surrounding conditions (for example, where airflow, contamination level, or water intrusion) are symmetrical. The condition may be nonzero if the surrounding conditions are nonsymmetrical. The concentration (C_s or C_v) at this point may be zero or nonzero for either symmetrical or nonsymmetrical conditions.

In the case of spore particles, the particles may go through metamorphosis and then migrate as vegetative fungal cells and cell collections. This migration may occur as a growing mold mat. Therefore, the mass balance in Figure 5.3 contains a reaction term for conversion of spores to vegetative cells. The vegetative cells then migrate through the paper. As they migrate, they digest paper, propagate, and form products, all expressed in the mass transfer mechanism shown in Figure 5.3. Since it is also possible for the vegetative cells to sporulate, the spores may then transfer back into the airstream as indicated. They may also transfer to the airstream as vegetative cells.

Each particle that adheres to a site becomes a collection of spores and source of migration and germination to vegetative growth. A multitude of growth centers are formed that spread out across the wallboard. In this fashion, the growth of the mold can be mathematically modeled and quantified. Similarly, the formation of secondary metabolites can be modeled and quantified. Eventually, we can predict the patterns we expect to see as we build the model to an expert system. The inocula to a wallboard can be spores or vegetative cells. This will of course alter the timing and degree of a mold infestation.

5.7.5 Spores to Mold Vegetative Cells, Back to Spores, and Renewal to Vegetative Cells: An Ongoing Process Unless Stopped

The formation of vegetative cells from spores and the reverse of forming spores from vegetative cells are based on complex reactions (the Ω terms in Figure 5.3). This is modeled based on assumed dependencies of conditions (moisture, temperature, nutrient availability, inhibitors, poisons, and perhaps potentiators). As vegetative cells migrate, they will propagate, thereby increasing in quantity, depending on available nutrients, oxygen, water, inhibitors, fungal poisons, and activators. The reaction rate terms (Ω) shown in Figure 5.3 are likely dependent on enzymes secreted by the mold to digest cellulose as an example.

5.7.6 The Role of Nutrition and Enzymes for Mold Propagation and Metamorphosis

The rate constants in the model will be dependent on the concentration and activity of these enzymes as well as the presence of inhibitors and activators. Since the enzyme activity will be a function of activators and inhibitors, the inherent mathematics become more complex. These will initially be assumed to follow a Monod (or modified Michaelis–Menten) dependency (Michaelis and Menten, 1913; Molz et al., 1986).

Sets of equations are considered that include (1) a reaction for conversion of spores to vegetative cells ($\Omega_{\text{spore transformation}}$); (2) one for conversion of vegetative cells to spores ($\Omega_{\text{vegetative transformation and growth}}$); (3) a model for spore hydration; (4) a reaction term for spore germination to vegetative growth (μ_s); (5) a model for enzyme formation and secretion (dE/dt) to enable the spore to access nutrition during germination; (6) equations for the kinetics of vegetative growth once spores have germinated (μ_v); (7) a model for enzyme formation and secretion (dE^1/dt) for enzyme formation and secretion to digest surroundings for nutrition for vegetative growth; and (8) an expression (μ_{v1}) for conversion of vegetative growth to spores.

The model assumed is that an air space to carry mold and moisture is adjacent to the top (paper) layer of the wallboard and another separate air space is adjacent to the bottom paper layer of the wallboard. There is mass transport between layers and between layers and air space above and below wallboard.

5.7.7 The Role of Moisture in Mold Propagation

Moisture content and movement are a function of water transport coupled with thermal effects as indicated in Figure 5.5. Moisture may move by diffusion, be affected by dispersion (D_w) and convection (U_w), and heat may be carried by water moving by diffusion, again affected by dispersion, or convection and also by thermally conductive transport (k). The heat and water transport may be radial along the surface of the wallboard as well as axial (into the wallboard) as shown. Water may evaporate or condense (Q_w). In doing so, heat is lost due to evaporative cooling that can act to cool the wallboard, or heat is gained by condensation thereby heating the wallboard

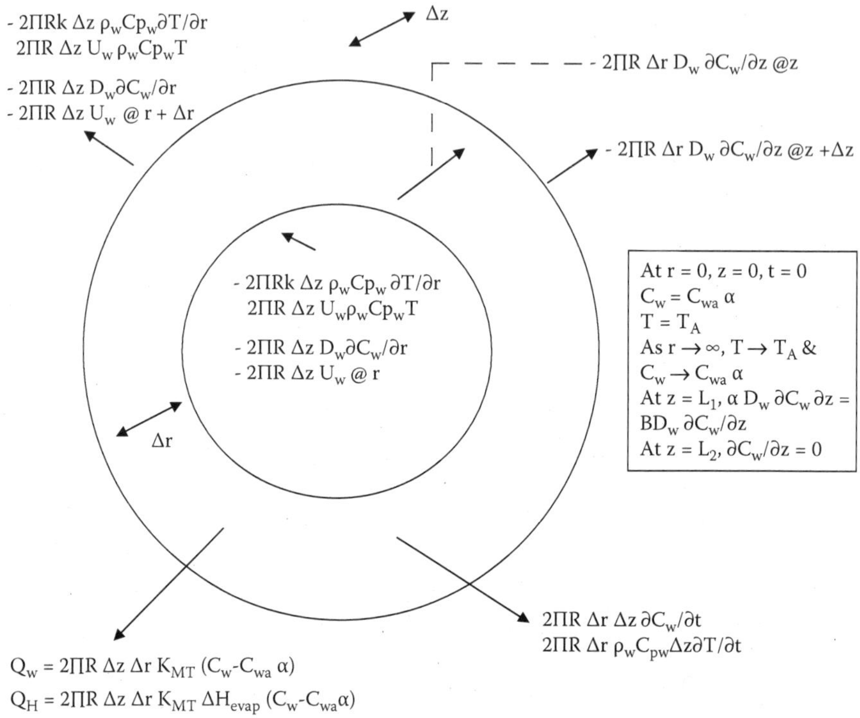

FIGURE 5.5 Shell balance pictorial of water and thermal transport in spore and vegetative growth space.

(term Q_H). Moisture is important since if moisture is absent, mold spores will not germinate, mold will not propagate in a vegetative manner, and mold products will not result. Moisture is never completely absent, so there are critical moisture levels that trigger spores to convert to vegetative cells. The moisture levels are temperature dependent.

Product and by-product formation are modeled as coupled phenomena dependent on mold state and health and conditions, including activator and inhibitors. Mold sporulation and conversion to vegetative growth (and vice versa) are critically dependent on moisture content and transport and also temperature. If ventilation and moisture were not a problem, then possibly mold infestation would not be a problem. The model accounts for radial and axial movement of mold products such as mycotoxins and secondary metabolites, plus for possible loss as a volatile substance.

The model accounts for the combination (Ω_{3sm}) of secondary metabolite formation (μ_{sm}) and destruction/inhibition (μ_d). Factors important in this are studied in computer experiments (in silico) and through analysis of experimental data from this program and available in the literature (see Figure 5.6).

All of these steps are analytically examined, representing points to possibly interfere with mold contamination. Thus, we envision possibilities for interferences in spore germination, sporulation, and release of secondary metabolites. Model molds

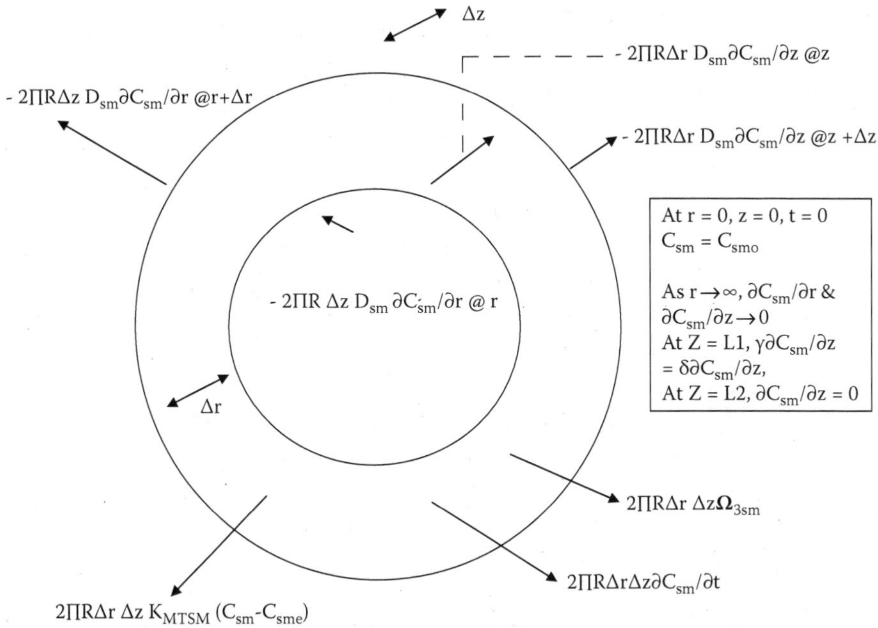

FIGURE 5.6 Shell balance pictorial of secondary metabolite transport in spore and vegetative growth space.

can be selected for examination in the laboratory that are likely to be found in field studies. The secondary metabolites are identified and the mold(s) genetically identified. Within these sequences, we can also begin to preliminarily look at gene segments that are relevant to formation of secondary metabolites that are dangers, and see how we may potentially interfere with such formation.

The literature contains information not specifically covering this model but possibly providing data and analysis that may be treated in the model or incorporated into it. For example, one paper uses data and correlation analyses to define the minimum humidity and respective temperature for a specific mold to propagate (Clarke et al., 1998). Another reference, a dissertation, provides useful background and data for input into the mathematical model for comparative purposes (Sedlbauer, 2002a). The reference (Sedlbauer, 2002b) develops a new hygrothermal calculation basis for analysis of buildings infested or potentially infested with mold. These references have other significant data, so assuming molds we identify match molds found or referenced in this paper, work proposed can be used to build upon and enhance the work of Sedlbauer (2002a, 2000b) and others.

Sedlbauer (2002a, 2000b) notes that the relationships involving spores and mycelia are complex. Using a fluorescence probe technique (mentioned by Sobek E.), one may be able to apply and use a specific assay method to distinguish spores from vegetative growth. This will of course allow one to distinguish spores from mycelia that may be mixed at a wallboard test surface so that appropriate data for use in the model can be obtained.

5.8 FORMULATION OF SYSTEM EQUATIONS FOR THE PAPER LAYER

Equations are formed into a set of partial differential equations by completing shell balances. These equations are converted to numerical format (as shown later in this chapter) for the formation and description of the numerical method. Each of the processes described are analyzed and graphs presented to illustrate trends for different variable sets and parameters.

5.8.1 Equation Description and Rationale for the Paper Layer

The equations for spore germination and vegetative growth and sporulation from vegetative biomass are presented first. This is a set of reaction equations that govern how products are biochemically formed. The reactions may possibly occur in the paper top layers of the wallboard and in the sublayer that is typically gypsum plaster since spores, cells, water, and heat transport may occur in all layers with communication in between.

5.8.2 Spores in the Paper Layer

Following Figure 5.3 relationships, we define

$$\Omega_{1\ \text{spore transformation}} = \mu_s C_s - \mu_v^1 C_v + \mu_{sd} C_s \tag{5.1}$$

$$\Omega_{2\ \text{vegetative transformation \& growth}} = -\mu_v C_v - \mu_s C_s + \mu_v^1 C_v + \mu_{vd} C_v \tag{5.2}$$

Equation 5.1 depicts that the reaction for spore transformation is composed of a reaction term for spore transformation to vegetative cells consisting of a reaction rate multiplied by the unsteady state spore concentration in the wallboard top surface at any point, $\mu_s C_s$. A term for the formation of spores from vegetative cells is subtracted, $\mu_v^1 C_v$, since spores are forming from vegetative cells on the right-hand side of the equation in Figure 5.3. Here, μ_v^1 is the overall rate of conversion of vegetative cells to spores, and C_v is vegetative cell concentration. A term is added for an overall rate of spore destruction as possible, given as $\mu_{sd} C_s$, where μ_{sd} is the rate of spore destruction per unit concentration of spores, and C_s is spore concentration. Similarly, Equation 5.2 is defined for vegetative cell transformation and growth. The first term is for the propagation of vegetative cells once formed ($-\mu_v C_v$). The next term if for the formation of vegetative cells from spores ($-\mu_s C_s$). The third term is for the conversion of vegetative cells back to spores, a loss ($\mu_v^1 C_v$). The last term, $\mu_{vd} C_v$, is for destruction of vegetative cells. This is confined to the very top and very base paper layers of the wallboard. The concentrations in these layers may of course differ. Within these paper layers, moisture intrusion is assumed to cause a spore particle to swell according to Equation 5.3.

5.8.2.1 Spore Hydration Kinetics

$$d(V_{\text{spore to cell}} C_w/dt) = D A_{\text{spore to cell}}(t)\,(C_{wex}\,\alpha - C_w) \tag{5.3}$$

where the volume $V_{\text{spore to cell}}$ of a spore expands as water enters, reaching concentrations, C_w (t), internally in the spore that is changing to a vegetative cell. Internal spore-to-vegetative cell water content rises with time to a set point that is metabolically controlled. Both $V_{\text{spore to cell}}(t)$ and $A_{\text{spore to cell}}(t)$ change since volume and area change.

5.8.2.2 Kinetics of Spore Germination to Vegetative Growth

$$\mu_s = (\mu_{s0})/(1 + k_{sw}/C_{sw1} + k_{s1}/N_{s1} + k_{s2}/N_{s2} + k_{is1} I_{is1} + \ldots) \tag{5.4}$$

N_{si} are nutrients and I_{isi} are germination inhibitors, where sets are $\Sigma_i\ k_{si}/N_{si}$ and $\Sigma_j\ k_{isj}N_{isj}$. The term k_{sw}/C_{sw1} is for water dependence. The k's are operating parameters analogous to K_m and K_i Michaelis–Menten constants. The term μ_{s0} is the basic reaction rate for spore germination to vegetative growth. Nutrient release must be controlled to begin to cease once sufficient nutrients are present. Such nutrients are then used up to support spore germination.

$$dN_{si}/dt = k_{nie} E_s/(1 + k_{ne}{*}E_s + k_i/N_{si} + k_{ii}N_{si}) - \mu_s\, \alpha_{Ni}\, C_s \tag{5.5}$$

N_{si} is a nutrient, where terms allow for nutrient inhibition or cutoff and control of reaction. The last term ($\mu_s\, \alpha_{Ni}\, C_s$) reflects stoichiometry in use of the nutrients to support spore germination. The k's are operating parameters analogous to K_m and K_i Michaelis–Menten constants, and E_s is the enzyme concentration necessary to cause release of nutrients. There may be several such enzymes specific to the particular nutrient released.

5.8.2.3 Enzyme Formation to Digest Surroundings for Nutrition for Spore Germination

$$dE_s/dt = \mu_{oe}\, C_s/(1 + k_e{*}E_s + k_{e1}/N_{s1} + k_{ei1}I_{ei1} + \ldots) \tag{5.5a}$$

where N_{ei} are nutrients, I_{eii} are enzyme inhibitors, and where sets are $\Sigma_i\ k_{ei}/N_{si}$ and $\Sigma_j\ ke_{ij}I_{eij}$. The k's are operating parameters analogous to K_m and K_i Michaelis–Menten constants.

For $\mu_{0e}\ C_s$, μ_{0e} is the basic reaction rate for enzyme formation to release nutrients to support spore germination to vegetative growth. The presence of spore concentration drives the formation of nutrients.

5.8.3 Vegetative Cells in the Paper Layer

5.8.3.1 Kinetics of Vegetative Growth Once Spores Have Germinated

$$\mu_v = (\mu_{v0})/(1 + k_{v1}/N_{v1} + \mathrm{k}_{v2}/N_{v2} + k_{iv1} I_{iv1} + \ldots) \tag{5.6}$$

N_{vi} are nutrients, where the set is $\Sigma_i\ k_{vi}/N_{vi}$ and I_{ivi} are vegetative growth inhibitors, where sets are $\Sigma_i\ k_{vi}/N_{vi}$ and $\Sigma_j\ k_{ivj}N_{ivj}$. The k's are operating parameters analogous to K_m and K_i Michaelis–Menten constants. For μ_{v0}, μ_{v0} is the basic reaction rate for vegetative growth.

Nutrient release must be controlled to begin to cease once sufficient nutrients are present. Such nutrients are then used up to support vegetative growth after spores have germinated.

$$dN_{vi}/dt = k_{nvie} E_v/(1 + k_{nve}{*}E_v + k_{vi}/N_{vi} + k_{vii}N_{vi}) - \mu_v\, \alpha_{Nvi}\, C_v E \tag{5.6a}$$

N_{vi} is a nutrient for vegetative growth, where terms allow for nutrient inhibition or cutoff and control of reaction. The last term ($\mu_v \alpha_{Nvi} C_v$) reflects stoichiometry in use of nutrients to support vegetative growth after spore germination The k's are operating parameters analogous to K_m and K_i Michaelis–Menten constants. An enzyme specific to the release of nutrients for vegetative growth is assumed. Note the enzyme may be inhibited as an intervention strategy as shown next.

5.8.3.2 Enzyme Formation and Secretion to Digest Surroundings for Nutrition for Vegetative Growth

$$dE_v/dt = \mu_{ove} C_v/(1 + k_{ev1}{}^* E_v + k_{ev2}/N_{ev2} + k_{eiv1}I_{eiv1} + \ldots) \tag{5.7}$$

where N_{evi} are nutrients, I_{eivi} are enzyme inhibitors, and where sets are $\sum_i k_{evi}/N_{evi}$ and $\sum_j k_{eivj}I_{eij}$. The k's are operating parameters analogous to K_m and K_i Michaelis–Menten constants.

For $\mu_{0ve} C_v$, μ_{0ve} is the basic reaction rate for enzyme formation to release nutrients to support spore germination to vegetative growth. The presence of the vegetative cell concentration drives the formation of nutrients for cell propagation.

5.8.3.3 Kinetics of Conversion of Vegetative Growth to Spores

$$\mu_v^1 = \mu_{vs0}^1/(1 + k_{vs1}/N_{vs1} + k_{vs2}/N_{vs2} + k_{ivs1} I_{ivs1} + k_w/C_w) \tag{5.8}$$

The last term accounts for modulation by water; low water triggers sporulation and high water content impedes sporulation. In the first case of low water content triggering sporulation, k_w should be in the vicinity of C_w. In the case of high water content, which should impede sporulation, again k_w can remain in vicinity of C_w. The term k_w can be postulated to change as a metabolic response to changes in water content. μ is the rate coefficient, k is a rate constant, E is enzyme concentration, N is an essential nutrient, and I is a key inhibitor or antagonist for the secondary metabolite. The nutrients and inhibitors noted may be special and control this process.

5.8.3.4 Vegetative Cell Dehydration Kinetics

The dehydration should be the reverse of the hydration process noted previously.

$$d(V_{\text{cell to spore}} C_w/dt) = D\, A_{\text{cell to spore}}(t)\,(C_{wex}\,\alpha - C_w) \tag{5.9}$$

where the volume $V_{\text{cell to spore}}$ of a cell decreases as water exits, reaching concentrations, $C_w(t)$, internally in the vegetative cell that is changing to a spore. Internal vegetative cell-to-spore water content decreases with time to a set point that is metabolically controlled. Both $V_{\text{cell to spore}}(t)$ and $A_{\text{cell to spore}}(t)$ change since volume and area change.

5.9 FORMULATION OF SYSTEM EQUATIONS FOR THE (GYPSUM) SUBLAYER

5.9.1 Spores in the Sublayer

$$\Omega_{1b\ \text{spore transformation}} = \mu_{sb} C_{sb} - \mu^1_{\underline{vb}} C_{vb} + \mu_{sbd} C_{sb} \tag{5.10}$$

$$\Omega_{2b\ \text{vegetative transformation \& growth}} = -\mu_{vb} C_{vb} - \mu_{sb} C_{sb} + \mu^1_{vb} C_{vb} + \mu_{vbd} C_{vb} \tag{5.11}$$

5.9.1.1 Spore Hydration Kinetics in the Sublayer

$$d(V_b\ C_{wb}/dt) = D_{sb}\ A_{\text{spore to cell}}\ (t)\ (C_{wexba}\ \alpha_{ab} - C_{wb}) \tag{5.12}$$

The volume of spore expands as water enters, reaching a certain concentration. Internal spore to vegetative cell water content rises with time to a set point. Attainment of that concentration will activate the enzyme trigger point. Moisture may be sourced from the paper layer above or air space below the sublayer, depending on the source of water.

Both V and A change since volume and area of spores changes as it is hydrated (or dried).

5.9.1.2 Kinetics of Spore Germination to Vegetative Growth in the Sublayer

$$\mu_{sb} = (\mu_{sb0})/\ (1 + k_{sbw}/C_{sbw1} + k_{sb1}/N_{sb1} + k_{sb2}/N_{sb2} + k_{isb1}\ I_{isb1} + \ldots) \tag{5.13}$$

N_{sbi} are nutrients and I_{isbi} are germination inhibitors in the sublayer, where sets are $\sum_i k_{sbi}/N_{sbi}$ and $\sum_j k_{isbj}N_{isbj}$. The term k_{sbw}/C_{sbw1} is for water dependence. The k's are operating parameters analogous to K_m and K_i Michaelis–Menten constants. For μ_{sb0}, μ_{sb0} is the basic reaction rate for spore germination to vegetative growth. Nutrient release must be controlled to begin to cease once sufficient nutrients are present. Such nutrients are then used up to support spore germination.

$$dN_{sbi}/dt = k_{nibe}\ E_{sb}/(1 + k_{neb}{}^*E_{sb} + k_i/N_{sbi} + k_{ibi}N_{sbi}) - \mu_{sb}\ \alpha_{Nbi}\ C_s \tag{5.14}$$

N_i is a nutrient, where terms allow for nutrient inhibition or cutoff and control of reaction. The second term reflects stoichiometry in use of nutrients to support spore germination

The k's are operating parameters analogous to K_m and K_i, Michaelis–Menten constants.

5.9.1.3 Enzyme Formation to Digest Surroundings for Nutrition for Spore Germination in the Gypsum Sublayer

$$dE_{sb}/dt = \mu_{oe}\ C_{sb}\ /\ (\ 1 + k_{eb}{}^*E_{sb} + k_{e1b}/N_{s1b} + k_{ei1b}I_{ei1b} + \ldots) \tag{5.15}$$

where N_{eib} are nutrients, I_{eiib} are enzyme inhibitors, and where sets are $\sum_i k_{eib}/N_{sib}$ and $\sum_j ke_{ijb}I_{eijb}$ in the sublayer. The k's are operating parameters analogous to K_m and K_i Michaelis–Menten constants. For $\mu_{0eb}\ C_{sb}$, μ_{0eb} is the basic reaction rate for enzyme formation to release nutrients to support spore germination to vegetative growth. The presence of spore concentration drives the formation of nutrients. This is again in the sublayer where nutrient insufficiency may be more of a possibility.

5.9.2 Vegetative Cells in the Sublayer

5.9.2.1 Kinetics of Vegetative Growth Once Spores Have Germinated in the Sublayer

$$\mu_{vb} = (\mu_{vb0})/(1 + k_{vb1}/N_{vb1} + k_{vb2}/N_{vb2} + k_{ivb1}\ I_{ivb1} + \ldots) \tag{5.16}$$

N_{vbi} are nutrients and I_{ivbi} are vegetative growth inhibitors, where sets are $\sum_i k_{vbi}/N_{vbi}$ and $\sum_j k_{ivbj}N_{ivbj}$ in the sublayer. The k's are operating parameters analogous to

K_m and K_i Michaelis–Menten constants. For μ_{vb0}, μ_{vb0} is the basic reaction rate for vegetative growth.

Nutrient release must be controlled to begin to cease once sufficient nutrients are present. Such nutrients are then used up to support vegetative growth after spores have germinated.

$$dN_{vb1}/dt = k_{nvieb}E_{vb} /(1 + k_{nveb}E_{vb} + k_{vb1}/N_{vb1} + k_{ivb1}N_{vb1}) - \mu_{vb}\, \alpha_{nvib}\, C_{vb} \tag{5.17}$$

N_{vib} is a nutrient for vegetative growth, where terms allow for nutrient inhibition or cutoff and control of reaction. The last term ($\mu_{vb}\, \alpha_{Nvib}\, C_{vb}$) reflects stoichiometry in use of nutrients to support vegetative growth after spore germination. The k's are operating parameters analogous to K_m and K_i Michaelis–Menten constants.

5.9.2.2 Enzyme Formation and Secretion to Digest Surroundings for Nutrition for Vegetative Growth in the Sublayer

$$dE_{vb}/dt = \mu_{oveb}\, C_{vb}/(1 + k_{evb1}E_{vb} + k_{evb2}/N_{evb2} + k_{eivb1}I_{eivb1} + \ldots) \tag{5.18}$$

where N_{evib} are nutrients, I_{eivib} are enzyme inhibitors, and where sets are $\sum_i k_{evib}/N_{evib}$ and $\sum_j k_{eivjb}I_{eijb}$. The k's are operating parameters analogous to K_m and K_i Michaelis–Menten constants.

For $\mu_{0veb}\, C_{vb}$, μ_{0veb} is the basic reaction rate for enzyme formation to release nutrients to support spore germination to vegetative growth. The presence of the vegetative cell concentration drives the formation of nutrients for cell propagation. Nutrient accessibility in the sublayer is required for this to occur.

5.9.2.3 Kinetics of Conversion of Vegetative Growth to Spores in the Sublayer

$$\mu^1{}_{vb} = \mu^1{}_{vsb0}/(1 + k_{vsb1}/N_{vsb1} + k_{vsb2}/N_{vsb2} + k_{ivsb1}\, I_{ivsb1} + k_{wb}/C_{wb}) \tag{5.19}$$

The last term accounts for modulation by water; low water triggers sporulation and high water should impede sporulation as noted for equations in the paper layer. The parameter, k_{wb}, will vary as water content varies. μ is the rate coefficient, k is a rate constant, E is enzyme concentration, N is an essential nutrient, and I is a key inhibitor or antagonist for the secondary metabolite.

The nutrients and inhibitors noted may be special and control this process. The dehydration should be the reverse of the hydration process noted previously. That is,

$$d(V_{\text{b cell to spore}}\, C_{wb}/dt) = D_b\, A_{\text{b cell to spore}}\, (t) * (C_{wb} - C_{wexb}\, \alpha_b) \tag{5.20}$$

both V and A change since volume and area change.

The sublayer (gypsum plaster for example) lies between top or frontal and bottom or backside layers (typically paper) of the wallboard. The reaction set for the top layer (Equations 5.1 to 5.9) applies equivalently to this layer with different boundary and initial conditions.

5.10 MASS BALANCES FOR SPORES AND VEGETATIVE CELLS IN REGARD TO THE PAPER LAYER

This section deals with equations that provide delivery of materials to reaction sites in the wallboard (top layers and sublayer) and removal of such materials.

5.10.1 Inlet Flow of Air Containing Spores Adjacent to the Top Layer of the Wallboard

Air may enter above and below (equivalently front and backside) of the wallboard perhaps at different rates and laden differently.

The mass balance for spores in the air space follows.

$$Q_{si}\, C_{si}\,(t) + K_{ms}\, A_{surf}\,(C_{ss}(t) - C_{sa}(t)) = Q_{si}C_{sa}(t) + V\partial C_{sa}/\partial t \qquad (5.21)$$

Spores are carried by volumetric flow. Net spores transported to and from the wallboard add to the flow. The term $K_{ms}\, A_{surf}\,(C_{ss}(t) - C_{sa}(t))$ may be negative or positive. Some spores transport back to the wallboard from the air space, some transport out of the volumetric space (V), and some may accumulate in the air space in this unsteady state process. The term $C_{ss}(t)$ must be given by an expression that allows the concentration at the surface to be different than the concentration in the air space.

Since the adsorption or binding of spores to the surface may be nonlinear, we assume that $C_{ss}(t)$ are related to $C_{sa}(t)$ by a form of an adsorption isotherm such that

$$C_{ss}(t) = V_{mads}\, C_{sa}\,(t)^{(1+\lambda_s)}/(K_{mads} + C_{sa}(t)) \qquad (5.21a)$$

In this expression, $C_{ss}(t) = C_{sa}(t)$ if $C_{sa}(t)$ is much larger than K_{mads}, and λ_s is zero and V_{mads} is one. This equation allows one to modulate adsorption at the surface by adjusting the equation to properties. It therefore is a modified Gibbs adsorption isotherm or Michaelis–Menten type expression (Chattori and Birdi, 1984; Michaelis and Menten, 1913). The term V_{mads} allows one to impose an extraction factor or proportional difference in association between the concentration of spores in the air space and the concentration of spores at the wallboard exterior surface.

5.10.2 Inlet Flow of Air Containing Vegetative Cells Adjacent to the Top Layer of the Wallboard

The mass balance for vegetative cells in the air space is as follows (interconversion of spores to vegetative cells and the reverse in the air space is not assumed though that is a possibility by occurring within a particle consisting of a mix of spores and vegetative cells perhaps).

$$Q_{vi}\, C_{vi}\,(t) + K_{mv}\, A_{surf}\,(C_{vs}(t) - C_{va}(t)) = Q_{vi}C_{va}(t) + V\partial C_{va}/\partial t \qquad (5.22)$$

Vegetative cells are carried in by volumetric flow. Net vegetative cells transported from the wallboard add to the flow. Therefore, the term $K_{mv}\, A_{surf}\,(C_{vss}(t) - C_{va}(t))$ may be positive or negative. Some vegetative cells transport back to the wallboard from the air space, some transport out of the volumetric space (V), and some may accumulate in the air space in this unsteady state process. The term $C_{vs}(t)$ must be given by an expression that allows the concentration at the surface to be different than the concentration in the air space.

As in the case of spores, since the adsorption or binding of vegetative cells to the surface may be nonlinear, we assume that $C_{vs}(t)$ are related to $C_{va}(t)$ by a form of an adsorption isotherm such that

$$C_{vs}(t) = V_{madv}\, C_{va}\,(t)^{(1+\lambda v)}/(K_{madv} + C_{va}(t)) \tag{5.22a}$$

In this expression, $C_{vs}(\mathrm{t}) = C_{va}(t)$ if $C_{sa}(t)$ is much larger than K_{madv}, and λ_v is zero and V_{madv} is one. Again, this equation allows one to modulate adsorption at the surface by adjusting the equation to properties. It therefore is a modified Gibbs adsorption isotherm or Michaelis–Menten type expression. The term V_{madv} allows one to impose an extraction factor or proportional difference in association between the concentration of vegetative cells in the air space and the concentration of vegetative cells at the wallboard exterior surface.

5.10.3 Mass Balance for Spores in the Top (Paper) Surface of the Wallboard

The mass balance for spores in the wallboard is provided next.

$$\begin{aligned} 0 = &-1/r\, \partial/\partial r(r\, D_{sr}\, \partial/\partial r(C_s)) + 1/r\, \partial/\partial r(r\, U_{sr}\, C_s) \\ &+ \partial C_s/\partial t - \partial/\partial z(D_{sz}\, \partial/\partial z\, (C_s)) + \partial(U_{sz}\, C_s)/\partial z + \Omega_{1\ spore} \end{aligned} \tag{5.23}$$

The term $\Omega_{1\ spore}$ reflects the net formed or lost due to reaction terms. The diffusive and convective mass transport terms in the r and z directions are included as is a term for accumulation of spores, $\partial C_s/\partial t$. The radial diffusion of spores in the wallboard plus the convective transport of spores radially in the wallboard exterior surface, plus the accumulation of spores in the circular segment of radial width Δr and thickness Δz, plus the diffusive transport of spores in the z dimension, plus the convective transport of spores in the z dimension, must equal the net rate of conversion of spores to vegetative cells and their destruction.

At $z = 0$, a boundary is imposed such that the mass transfer of spores from the air space at the frontal surface of the wallboard equals the diffusive and convective flux into the wallboard in the z direction.

$$K_{ms}\, A_s\, (C_{sa} - C_{ss}) = -A_s\, D_{sz}\, \partial/\partial z(C_s) + U_{sz}\, A_s\, C_s \text{ at } z = 0 \tag{5.23a}$$

The term C_{ss} is related to the term C_{sa} by the adsorption term in Equation 5.21a as well as by the mass balance and reaction contribution at $z = 0$ from Equations 5.23 and 5.23a.

5.10.4 Mass Balance for Vegetative Cells in the Wallboard Paper Layer

$$0 = -1/r\ \partial/\partial r(r\ D_{vr}\ \partial/\partial r\ (C_v)) + 1/r\ \partial/\partial r(r\ U_r\ C_v)$$
$$+ \partial C_v/\partial t - \partial/\partial z(D_{vz}\ \partial/\partial z\ (C_v)) + \partial(U_{vz}\ C_s)/\partial z$$
$$+ \Omega_{2\ \text{vegetative transformation \& growth}} \tag{5.24}$$

At $z = 0$, a boundary is imposed such that the mass transfer of vegetative cells from the air space at the frontal surface of the wallboard equals the diffusive and convective flux into the wallboard in the z direction.

$$K_{mv} A_s(C_{va} - C_{vs}) = -A_s\ D_{vz}\ \partial/\partial z\ (C_v) + A_s\ \partial(U_{vz}\ C_s)/\partial z \tag{5.24a}$$

The term C_{vs} is related to the term C_{va} by the adsorption term in Equation 5.22a.

Similar to Equation 5.23 for spores, in this case for vegetative cells, the terms in equation 5.24a reflect what is transported, and the net C_v formed or lost due to reaction term $\Omega_{2\ \text{vegetative transformation \& growth}}$. The term C_{vs} is also governed by its relationship to C_{va} via Equation 5.22a. The term C_{va} is governed by the mass transfer equation in the air space.

Therefore, in Equation 5.24, the radial diffusion of vegetative cells in the wallboard plus the convective radial transport of vegetative cells in the surface of the wallboard, plus the accumulation of vegetative cells in the circular segment of radial width Δr and thickness Δz, plus the diffusive transport of vegetative cells in the z dimension, plus the convective transport of vegetative cells in the z dimension must equal the net rate of conversion of vegetative cells to spores and their destruction.

5.11 MASS BALANCE FOR SPORES AND VEGETATIVE CELLS IN THE AIR SPACE ADJACENT TO THE BOTTOM OR EQUIVALENTLY BACK SURFACE OF THE WALLBOARD

The airflow over the bottom or back surface may differ from that over the top surface. For example, it may be hidden between walls and consist only of minimal natural convection. If the flow is small enough such as not to interact with the wallboard, then the bottom or back paper surface may simply be where transport of spores or vegetative cells is zero at this surface. The spores and vegetative cells may simply accumulate or be depleted due to transport affecting the top or frontal surface.

5.11.1 Inlet Flow of Air Containing Spores Adjacent to the Backside Paper Layer of the Wallboard

The mass balance for spores in the air space bordering the back or bottom surface follows.

$$Q_{sib}\ C_{sib}\ (t) + K_{msb}\ A_{surfb}\ (C_{sbs}(t) - C_{sab}(t)) = Q_{sib} C_{sab}(t) + V_b\ \partial C_{sab}/\partial t \tag{5.25}$$

Spore flow is carried in by volumetric flow that exists below the bottom (or equivalently bathing the backside surface). The net rate of spores transported from the wallboard outer surface add to the flow. Some sores transport out of the volumetric air space (V_b), and some may accumulate in the air space in this unsteady state process. The surface spore concentration is also related to the air space concentration is the manner given in Equation 5.21a.

5.11.2 Inlet Flow of Air Containing Vegetative Cells Adjacent to the Backside Paper Layer of the Wallboard

The mass balance for vegetative cells in the backside air space is as follows (interconversion of spores to vegetative cells and the reverse in the air space is not assumed though that is a possibility).

$$Q_{vib}\, C_{vib}\,(t) + K_{mvb}\, A_{surfb}\, (C_{vbs}(t) - C_{vab}(t)) = Q_{vib} C_{vab}(t) + V_b \partial C_{vab}/\partial t \qquad (5.26)$$

Vegetative cells flow into the wallboard backside or bottom side air space carried by volumetric flow. Vegetative cells transported from the wallboard add to the air flow. Some vegetative cells transport out of the volumetric space (V_b), and some may accumulate in the air space in this unsteady state process. Equations 5.21a and 5.22a for the adsorption type relationship between air space and surface apply for the backside surface as well for spores and vegetative cells, respectively.

5.11.3 Mass Balance for Spores in the Backside (Paper) Surface of the Wallboard

$$0 = -1/r\, \partial/\partial r\, (r\, D_{srb}\, \partial/\partial r\, (C_{sb})) - U_{srb}/r\, \partial/\partial r\, (r\, C_{sb})$$
$$+ \partial C_{sb}/\partial t + \partial/\partial z\, (D_{szb}\, \partial/\partial z\, (C_{sb})) - U_{szb}\, \partial C_{sb}/\partial z + \Omega_{1b\ spore} \qquad (5.27)$$

At $z = 1 + 2p$, a boundary is imposed such that the mass transfer of spores from the air space at the backside surface of the wallboard equals the diffusive and convective flux into the wallboard in the z direction.

The term C_{vs} is related to the term C_{va} by the adsorption term in Equation 5.22a.

$$K_{msb}\, (C_{sab} - C_{sbs})\, A_b = A_b\, D_{szb}\, \partial/\partial z\, (C_{sb}) - A_b\, U_{szb}\, \partial C_{sb}/\partial z \qquad (5.27a)$$

at $z = Z_{2p+L}$.

The term C_{sbs} is related to the term C_{sab} by the adsorption term in equation 5.21a.

The radial diffusion of spores in the wallboard backside outer layer plus the radially convective transport of spores in the wallboard surface layer, plus the accumulation of spores in the circular segment of radial width Δr and thickness Δz, plus the diffusive transport of spores in the z dimension, plus the convective transport of spores in the z dimension is equal to the net rate of conversion of spores to vegetative cells and their destruction. Again, it is assumed that the spores at the surface follow the concentration in the air space according to an adsorption phenomena given in

Equation 5.21a. The spores (and vegetative cells) at the surface are therefore always in a dynamic equilibrium with those in the air space.

5.11.4 Boundary Condition between the Backside Paper Layer and the Gypsum Sublayer in Regard to Spores

A boundary condition also exists between the bottom (backside) paper layer and the gypsum sublayer for spores such that

$$-D_{sz}\,\partial C_s/\partial z + U_{sz}\,C_s = -D_{szb}\,\partial C_{sb}/\partial z + U_{szb}\,C_{sb} \tag{5.28}$$

where $z = z_{p+1}$ noting that the median of the sublayer surface is at $z = z_1$. The top of paper layer is at $z = 0$, and the top or frontal layer ends at $z = z_p$. The gypsum sublayer starts at $z = z_p$ and ends at $z = z_{p+1}$ (see Figure 5.4).

5.11.5 Mass Balance for Vegetative Cells in the Backside Wallboard Paper Layer

The mass balance for vegetative cells in the (backside) surface layer follows.

$$0 = -1/r\,\partial/\partial r\,(r\,\mathrm{D}_{vrb}\,\partial/\partial r\,(C_{vb})) + 1/r\,\partial/\partial r\,(r\,U_{vrb}\,C_{vb})$$
$$+\,\partial C_{vb}/\partial t - \partial/\partial z\,(D_{vbz}\,\partial\,C_{vb}\,/\partial z) + U_{vzb}\,\partial C_{sb}\,/\partial z + \Omega_{1b\ vegetative} \tag{5.29}$$

At $z = 1 + 2p$, a boundary is imposed such that the mass transfer of vegetative cells from the air space at the backside surface of the wallboard equals the diffusive and convective flux into the wallboard in the z direction.

The term C_{vs} is related to the term C_{va} by the adsorption term in Equation 5.22a.

$$K_{mvb}\,A_b(C_{vab} - C_{vbs}) = -D_{vbz}A_b\,\partial C_{vb}\,/\partial z + A_b\,U_{vzb}\,C_{sb} \tag{5.29a}$$

at $z = z_{2p+L}$.

The radial diffusion of vegetative cells in the wallboard back surface plus the radial convective transport of vegetative cells in the wallboard back surface, plus the accumulation of vegetative cells in the circular segment of radial width Δr and thickness Δz in the back surface paper layer, plus the diffusive transport of vegetative cells in the z dimension, plus the convective transport of vegetative cells in the z dimension is balanced by the net rate of conversion of vegetative cells to spores and their destruction in the back surface layer. The vegetative cells in the air space are assumed to be in dynamic equilibrium with those in the backside wall board surface though allowed to change with time.

5.11.6 Boundary Condition between the Backside Paper Layer and the Gypsum Sublayer in Regard to Vegetative Cells

A boundary condition also exists between the back surface paper layer and the gypsum sublayer for vegetative cells such that

$$-D_{vz}\,\partial C_v/\partial z + U_{vz}C_s = -D_{vzb}\,\partial C_{vb}/\partial z + U_{vzb}\,C_{vb} \tag{5.30}$$

where $z = z_{p+1}$ noting that the median of the sublayer surface is at $z = z_1$. The top of the paper layer is at $z = 0$, and the top or frontal layer ends at $z = z_p$. The gypsum sublayer starts at $z = z_p$ and ends at $z = z_{p+1}$ (see Figure 5.4).

5.12 WATER TRANSPORT

Water enters or leaves the air space adjacent to the top (front) layer and adjacent to the bottom (backside) layer of the wallboard and through boundaries and by exchange with the wallboard. Water in the air space enters and leaves by convection by assumption. Water may of course enter as a liquid from, for example, a leaking pipe and saturate the wallboard. The expression for water content mass exchange in the air space in contact with the wallboard follows.

5.12.1 Mass Balance for Water in the Air Space Adjacent to the Paper Layer

$$Q_a\, C_{wi}(t) + K_w\, A_s(C_w - C_{wa}(t)\, \beta) = Q_a C_{wa}(t) + V\partial C_{wa}/\partial t \tag{5.31}$$

In this expression air enters the space at volumetric rate (Q_a) with water content ($C_{wi}(t)$) allowed to vary with time. The net amount of water evaporating from the wallboard surface adds to the air space at a rate governed by a mass transfer coefficient (K_w), wallboard surface area (A_s), and the driving force of water concentration in the wallboard (C_w) less the water concentration in the air space adjusted for the absorptivity (β) of the water in the wallboard ($C_{wa}(t)\ \beta$). This term may of course be negative or positive depending on whether water is by net rate entering or leaving the air space adjacent to the paper layers.

5.12.2 Mass Balance for Water in the Top or Bottom (Paper) Layer of the Wallboard

The water content of air and wallboard depends on temperature since water will evaporate or condense depending on temperature. Water migrates through the paper top layer and the sublayer, typically the gypsum layer, exchanging with surroundings. The water content of air and wallboard depends on temperature since water will evaporate or condense depending on temperature.

5.12.3 Mass Balance for Water Transport in the Paper Top Layer

For water transport in the paper top layer,

$$\begin{aligned} 0 = {} & -1/r\, \partial/\partial r(r\, D_{wr}\, \partial C_w/\partial r) + 1/r\, \partial/\partial r(r\, U_{wr}\, C_w) \\ & + \partial C_w/\partial t - \partial/\partial z(D_{wz}\, \partial C_w/\partial z) + \partial(U_{wz}\, C_w)/\partial z \end{aligned} \tag{5.32}$$

where the boundary condition at $z = 0$ is

$$K_w\, A_s(C_{wa}\, \beta - C_w) = -D_{wz}\, \partial C_w/\partial z + \partial\, (U_{wz}\, C_w)/\partial z \tag{5.32a}$$

5.12.4 Mass Balance for Water in the Air Space Adjacent to the Backside Paper Layer of the Wallboard Surface

For transport of water in the air space adjacent to the backside paper layer, water enters or leaves the air space adjacent to the backside layer of the wallboard through boundaries and by exchange with the backside paper layer of the wallboard. Water in the air space enters and leaves by convection by assumption. The expression for water content mass exchange in the air space in contact with the wallboard backside paper layer follows.

$$Q_{ab}\, C_{wbi}(t) + K_{wb}\, A_{surfb}\, (C_{wwb} - C_{wab}(t)\ \beta_b) = Q_{ab} C_{wab}(t) + V_b \partial C_{wab}/\partial t \qquad (5.33)$$

In this expression air enters the air space adjacent to the sublayer of the wallboard at volumetric rate (Q_{ab}) with water content ($C_{wbi}(t)$), allowed to vary with time. The net rate of water evaporating from the sublayer wallboard surface adds to the air space at a rate governed by a mass transfer coefficient (K_{wb}), wallboard surface area (A_{surfb}), and the driving force of water concentration in the wallboard (C_{wwb}) less the water concentration in the air space adjusted for the absorptivity (β_b) of the water in the wallboard ($C_{wab}(t)\ \beta_b$). The terms C_{wwb} and C_{wb} (the latter in Equation 5.34) are the same entities in the context of the model though differing in units for the sake of numerical calculation.

5.12.5 Mass Balance for Water in the Backside Paper Layer

For water transport in the backside paper layer,

$$\begin{aligned} 0 = &-1/r\, \partial/\partial r(r\, D_{wrb}\, \partial C_{wb}/\partial r) + 1/r\, \partial/\partial r(r\, U_{wrb}\, C_{wb}) \\ &+ \partial C_{wb}/\partial t - \partial/\partial z\, (D_{wzb}\, \partial C_{wb}/\partial z) + \partial(U_{wbz}\, C_{wb})/\partial z \end{aligned} \qquad (5.34)$$

where the boundary condition

$$K_{wb}\, A_{surfb}\, (C_{wba}\ \beta_b - C_{wwb}) = -D_{wzb}\, \partial C_{wb}/\partial z + U_{wbz}\, C_{wb} \qquad (5.34a)$$

exists at $z = z_{2p+L}$.

5.12.6 Flux Boundary Condition for Water between the Top or Bottom Paper Layer on the Wallboard and the Sublayer (Gypsum)

At the interface between the top paper layer on the wallboard and the sublayer (gypsum) below the top layer, as well at the interface between the backside paper layer and the gypsum sublayer, a flux boundary condition exists for water transport, that is,

$$-D_{wz}\, \partial C_w/\partial z + \partial(U_{wz}\, C_w)/\partial z = -D_{wzb}\, \partial C_{wb}/\partial z + \partial(U_{wzb}\, C_{wb})/\partial z \qquad (5.35)$$

where the boundaries are at $z = z_p$ and at $z = z_{p+1}$ for this equation applied at each of these positions.

5.12.7 Mass Balance for Water in the Gypsum Sublayer

Water transfer in the gypsum sublayer occurs by diffusion and possibly convection. Water may evaporate from the paper surfaces (or condense onto the surfaces) where the surfaces are adjacent to the gypsum layer. It is assumed that such evaporation is a boundary condition that may occur at the frontal and backside surfaces of the gypsum layer. Accordingly, for the gypsum layer,

$$0 = -1/r\, \partial/\partial r(r\, D_{wrb}\, \partial C_{wb}/\partial r) + 1/r\, \partial/\partial r\,(r\, U_{wrb}\, C_{wb})$$
$$+ \partial C_{wb}/\partial t - \partial/\partial z(D_{wzb}\, \partial C_{wb}/\partial z) + \partial(U_{wzb}\, C_{wb})/\partial z \qquad (5.35a)$$

5.13 THERMAL TRANSPORT

5.13.1 Thermal Transport in the Air Space Adjacent to the Top (Paper) Layer of the Wallboard

Water transport is linked to temperature so first deal with temperature in the air space adjacent to the frontal layer of the wallboard.

$$Q_a\, \rho_i C_{pi}\, T_i(t) + Q_{ht} + h_e\, A_{surf}\,(10^4\ \mathrm{cm^2/m^2})(T_s(t) - T_e(t))$$
$$+ K_w\, A_{surf}\,(C_{ww} - C_{wa}\,\alpha)\, \Delta H_{evap} = Q_{ae}\, \rho_e C_{pe}\, T_e(t)$$
$$+ \partial\,(\rho_e C_{pe}\, V\, T_e)/\partial t \qquad (5.36)$$

where $\rho_e = 0.354\, T_e$ with air as a perfect gas.

Equation 5.36 first consists of heat convectively carried in which is enthalpy (equal to $Q_a\, \rho_i C_{pi}\, T_i(t)$), where $\rho_i C_{pi}\, T_i(t)$ consists of air density, specific heat, and temperature in the inlet. The second term, Q_{ht}, allows for heating (or cooling of the air space, which would then be a negative second term). The third term, $h_e\, A_{surf}\,(10^4\ \mathrm{cm^2/m^2})\ (T_s(t) - T_e(t))$, accounts for net convective heat transfer to the wallboard frontal surface. The fourth term, $K_w\, A_{surf}\,(C_{ww} - C_{wa}\,\alpha)\, \Delta H_{evap}$, accounts for heating of the air due to evaporation from the frontal wallboard surface. The fourth term, $Q_{ae}\, \rho_e C_{pe}\, T_e(t)$, accounts for enthalpic heat removed from the air space. The last term, $\partial(\rho_e C_{pe}\, V\, T_e)/\partial t$, accounts for unsteady temperature change in the air space adjacent to the wallboard.

5.13.2 Thermal Transport in the Frontal and Backside (Paper) Layers of the Wallboard

The thermal balance in the top or backside surface layer of the wallboard is

$$0 = -1/r\, \partial/\partial r\,(r\, k_{wrs}\, \partial/\partial r\,(T)) + 1/r\, \partial/\partial r\,(r\, \rho_{wrs}\, C_{pwrs}\, U_{wrs}\, T) - \partial/\partial z\,(k_{wzs}\, \partial/\partial z\,(T))$$
$$+ \partial\,(\rho_{wzs}\, C_{pwzs}\, U_{wzs}\, T)/\partial z + \rho_{ws}\, C_{pws}\, \partial T/\partial t \qquad (5.37)$$

In the equation, $1/r\, \partial/\partial r\,(r\, \rho_{wrs}\, C_{pwrs}\, k_{wrs}\, \partial/\partial r\,(T))$ is for conductive radial transfer in the frontal surface layer, and $1/r\, \partial/\partial r\,(r\, \rho_{wrs}\, C_{pwrs}\, U_{wrs}\, T)$ is for convective heat transfer radially in the frontal surface layer. The next two terms, $-\partial/\partial z\ (k_{wzs}\ \partial/\partial z\ (T)) + \partial\ (\rho_{wzs}\ C_{pwzs}\ U_{wzs}\ T)/\partial z$, cover conductive and convective heat transfer into the depth of the paper (for completeness only, since it is expectedly quite thin). The last term, $\rho_{ws}\ C_{pws}\ \partial T/\partial t$, covers unsteady state heating of the top layer.

At $z = 0$, the following boundary condition exists:

$$h_e A_{surf}(T_e - T) = K_w A_{surf}\,(C_w - C_{wa}\,\alpha)\,\Delta H_{evap\ surf} - k_{wzs}\,A_{surf}\partial/\partial z\,(T) + \rho_{wzs}\,C_{pwzs}\,A_{surf}\,U_{wzs}\,T \tag{5.37a}$$

In Equation 5.37a, $h_e A_{surf}\,(T_e - T)$ is for net convective heat added to the frontal wallboard surface from the air where h_e is the heat transfer per unit depth which should be the same as heat transfer per unit surface. The next term, $K_w\,A_{surf}\,(C_w - C_{wa}\,\alpha)\,\Delta H_{evap\ surf}$, covers heat loss due to evaporative cooling from the top layer at $z = 0$. The last term, $\rho_{ws}\,C_{pws}\,A_{surf}\,U_{wzs}\,T$, covers the possibility of convective removed in the z direction at $z = 0$.

5.13.3 Thermal Transport in the Gypsum Sublayer of the Wallboard

Similarly for the gypsum layer of the wallboard, the following equations apply for the thermal balance.

The thermal balance in the gypsum sublayer of the wallboard (bounded by the surface paper layers) is

$$0 = -1/r\,\partial/\partial r(r\,\rho_{wrsl}\,C_{pwrsl}k_{wrsl}\,\partial/\partial r(T_{sl})) + 1/r\,\partial/\partial r(r\,\rho_{wrsl}\,C_{pwrsl}\,U_{wrsl}\,T_{sl}) - \partial/\partial z\,(\rho_{wzsl}\,C_{pwzsl}\,k_{wzsl}\,\partial/\partial z(T_{sl})) + \partial(\rho_{wzsl}\,C_{pwzsl}\,U_{wzsl}\,T_{sl})/\partial z + \rho_{wsl}\,C_{pwsl}\,\partial T_{sl}/\partial t \tag{5.38}$$

Heat will be added to (or lost from) the gypsum sublayer at the boundaries with the paper layer by convection, conduction, or by thermal evaporation as a boundary condition at $z = z_p$ and $z = z_{p+1}$. The term $1/r\,\partial/\partial r\,(r\,\rho_{wrsl}\,C_{pwrsl}\,k_{wrsl}\,\partial/\partial r\,(T_{sl}))$ is for conductive heat transfer in the gypsum sublayer, and $1/r\,\partial/\partial r\,(r\,\rho_{wrsl}\,C_{pwrsl}\,U_{wrsl}\,T)$ is for convective heat transfer radially in the gypsum sublayer. The next two terms, $-\partial/\partial z\,(\rho_{wzsl}\,C_{pwzsl}\,k_{wzsl}\,\partial/\partial z\,(T_{sl})) + -\partial\,(\rho_{wzsl}\,C_{pwzsl}\,U_{wzsl}\,T_{sl})/\partial z$, cover conductive and convective heat transfer into the depth of the paper in the gypsum sublayer. The last term, $\rho_{wsl}\,C_{pwsl}\,\partial T_{sl}/\partial t$, covers unsteady state heating of the gypsum sublayer.

5.13.4 Heat Exchange between the Top (Backside) Layers and the Gypsum Sublayer

The expression for heat exchange between the top (backside) layers and the gypsum sublayer of the wallboard is

$$-\rho_{wzs}\,C_{pwzs}\,k_{wzs}\,\partial\,T/\partial z + \rho_{wzs}\,C_{pwzs}\,U_{wzs}\,T) + k_{wzs}\,(C_{wzs} - C_{wzb}\,\alpha_{sl})\,\Delta H_{\text{evap sublayer}} = -\rho_{wzb}\,C_{pwzb}\,k_{wzn}\,\partial T_{sl}/\partial z + U_{wzb}\,T_{sl}) \tag{5.39}$$

where $z = z_p$ or $z = z_{p+l}$. This accounts for conductive heat exchange, convective heat exchange, and the possible passage of water carrying heat or providing cooling across the interfaces.

5.13.5 Thermal Transport in the Air Space Adjacent to the Backside Paper Layer of the Wallboard

$$Q_{ab}\,\rho_{ib}C_{pib}\,T_{ib}(t) + Q_{htb} + h_{eb}\,A_{sb}(T_{eb}(t) - T_{sb}(t)) + k_{swb}\,A_{sb}(C_{wb} - C_{wab}\,\alpha_b)\Delta H_{\text{evapsublayerairspace}}$$
$$= Q_{aeb}\,\rho_{eb}C_{peb}\,T_{eb}(t) + \partial(\rho_{eb}\,C_{peb}\,V_{eb}\,T_{eb})/\partial t \qquad (5.40)$$

This equation first consists of heat convectively carried in which is enthalpy (equal to $Q_{ab}\,\rho_{ib}C_{pib}T_{ib}(t)$), where $\rho_{ib}C_{pib}\,T_{ib}(t)$ consists of air density, specific heat and temperature in the inlet. The second term, Q_{htb}, allows for heating (or cooling of the air space, which is a negative second term). The third term, $h_{eb}\,A_{sb}\,(T_{eb}(t) - T_{sb}(t))$, accounts for convective heat transfer to the wallboard sublayer surface. The fourth term, $k_{swb}\,A_{sb}\,(C_{wb} - C_{wab}\,\alpha_b)\,\Delta H_{\text{evapsublayerairspace}}$, accounts for heating of the air due to condensation of water evaporated from the wallboard sublayer surface. The term $Q_{aeb}\,\rho_{eb}C_{peb}\,T_{eb}(t)$ accounts for enthalpic heat removed from the sublayer air space. The last term, $\partial(\rho_{eb}C_{peb}\,V_{eb}\,T_{eb})/\partial t$, accounts for unsteady temperature change in the air space above the wallboard.

5.14 RELATIONSHIP BETWEEN AIR WATER CONTENT AND TEMPERATURE IN AIR SPACE

The relationship between temperature and moisture content in the air space above and below the wallboard is as provided in the literature where the saturated water content is a function of temperature. The water content at a particular fraction of saturation is also a function of temperature. This temperature is referred to as the dry bulb temperature.

5.15 EQUATION FOR WEIGHT OF WATER CONTENT PER WEIGHT OF DRY AIR

An equation for weight of water content per weight of dry air follows (Tetens, 1930)

$$\text{kg water/kg dry air} = 3.7868 \times 10^{-3} \exp\,(17.2694\,t_c/[t_c + 238.3]) \qquad (5.40a)$$

At the same temperature the fraction of saturation or relative humidity is simply multiplied by equation 5.40a to obtain the water content in the air.

Moisture content in the wallboard paper and gypsum plaster subsurface is set at an initial value depending on conditions. It will saturate to a maximal extent. Water may soak the wallboard due to a liquid intrusion or be adsorbed from the air space depending on entry conditions. The water adsorption may follow the form of adsorption of spores or cells in a mathematical pattern. However, for our purposes, the water content will be set by intrusion to a maximal value. In the case of nonliquid intrusion but exposure from the air space, it will simply be assumed to follow the air space value at the surface according to a proportionality α for the air space frontal layer boundary. It is possible that such a proportional relationship may not exist so an alternative is to consider water adsorption following water content in the air space to replace the term α according to the following.

Since the adsorption or binding of water to the surface may be nonlinear, we assume that $C_w(t)$ and $C_{wa}(t)$ are related by a form of an adsorption isotherm at equilibrium such that

$$C_w(t) = \alpha_{wak}\, C_{wa}(t)^{(1+\lambda ws)}/(K_{wads} + C_{wa}(t)) \tag{5.40b}$$

In this expression, $C_w(t) = \beta C_{wa}(t)$ if K_{wads} is much larger than $C_{wa}(t)$ and λ_{ws} is zero, where $\beta = \alpha_{wak}/K_{wads}$. This equation allows one to modulate adsorption at the surface by adjusting the equation to properties. It therefore is a modified Gibbs adsorption isotherm or Michaelis–Menten type expression (Chattoraj and Birdi, 1984; Michaelis and Menten, 1913). The term α allows one to impose an extraction factor or proportional difference in association between the concentration of water in the air space and the concentration of water at the wallboard exterior surface.

That completes the equation set except for the important aspect of by-products or secondary or near primary metabolites of mold that have health implications. The boundary and initial conditions will depend on the case being investigated.

5.16 SECONDARY METABOLITE MASS BALANCES AND REACTIONS

5.16.1 Secondary Metabolite Mass Balances and Reactions for the Air Space Adjacent to the Paper Layer of the Wallboard

The mass balance for metabolites produced by vegetative cells and appearing in the air space follows. The spores are seedlike and not expected to be productive entities so metabolites are not expected to be produced by them in any degree until they vegetatively propagate.

$$\begin{aligned} &Q_{mai}\,(C_{smai}(t) + x_{mai}(t)\,C_{sai}(t) + y_{vmai}(t)\,C_{vai}(t)) + k_{mtsm}\,A_{surf}\,(C_{sm}(t) - C_{sma}(t)) \\ &\quad + k_{mtsms}\,A_{surf}\,(x_{ssm}(t)\,C_s(t) - x_{ssma}(t)\,C_{sa}(t)) + k_{mtsmv}\,A_{surf}\,(y_{vsm}(t)\,C_v(t) \\ &\quad - y_{vsma}(t)\,C_{va}(t)) = Q_{ma}\,(C_{sma}(t) + x_{ma}(t)\,C_{sa}(t) + y_{vma}(t)\,C_{va}(t)) + V\partial(C_{sm}(t) \\ &\quad + x_{ssm}(t)\,C_s(t) + y_{vsm}(t)\,C_v(t))/\partial t \end{aligned} \tag{5.41}$$

Volumetric flow carries in free metabolites produced by the cells, $C_{smai}(t)$, metabolites imbedded in the spore particles, $x_{mai}(t)\,C_{sai}(t)$, and metabolites imbedded in the vegetative cells, $y_{vmai}(t)\,C_{vai}(t)$, as the first term. The grouping of three transport sets consists of, respectively, net secondary metabolites transported from the surface as free entities, $k_{mtsm}\,A_{surf}\,(C_{sm}(t) - C_{sma}(t))$, entities contained in spores, $k_{mtsms}\,A_{surf}$ $(x_{ssm}(t)\,C_s(t) - x_{ssma}(t)\,C_{sa}(t))$, and entities contained in vegetative cells, $k_{mtsmv}\,A_{surf}$ $(y_{vsm}(t)\,C_v(t) - y_{vsmai}(t)\,C_{vai}(t))$. Notice that the transport terms are separated since free secondary metabolites, those in spore particles and those in vegetative cell particles may transport differently between the top surface and the air space adjacent to the top surface of the wallboard. Some secondary metabolites transport out of the volumetric space (V), Q_{ma} $(C_{sma}(t) + x_{ma}(t)\, C_{sa}(t) + y_{vma}(t)\, C_{va}(t))$, and some may accumulate in the air space in this unsteady state process, $V\partial(C_{sm}(t) + x_{ssm}(t)\, C_s(t) + y_{vsm}(t)\, C_v(t))/\partial t$.

5.16.2 Secondary Metabolite Balances for the Paper Layers of the Wallboard

The mass balance for secondary metabolites in the top or backside surfaces of the wallboard is provided next. These may be freely moving, or may be moving as part of spores or as part of vegetative cells as indicated.

$$\begin{aligned}\Omega_{3sm} = & -1/r\, \partial/\partial r\, (r\, D_{smr}\, \partial/\partial r(C_{sm}(t) + r\, D_{sr}\, \partial/\partial r\, (x_{ssm}(t)\, C_s(t)) \\ & + r\, D_{vr}\, \partial/\partial r\, (y_{vsm}(t)\, C_v(t))) + 1/r\, \partial/\partial r\, (r\, U_{smr}\, (C_{sm}(t) + r\, U_{sr}\, x_{ssm}(t)\, C_s(t) \\ & + r\, U_{vr}\, y_{vsm}(t)\, C_v(t)) + \partial((C_{sm}(t) + x_{ssm}(t)\, C_s(t) + y_{vsm}(t)\, C_v(t))/\partial t \\ & + \partial/\partial z\, (D_{smz}\, \partial C_{sm}(t)/\partial z + D_{sz}\, \partial\, (x_{ssm}(t)\, C_s(t))/\, \partial z \\ & + D_{vz}\, \partial\, (y_{vsm}(t)\, C_v(t))/\, \partial z) - \partial\, (U_{smz}\, C_{sm}(t) + U_{sz}\, x_{ssm}(t)\, C_s(t) \\ & + U_{vz}\, y_{vsm}(t)\, C_v(t))\, /\partial z\end{aligned} \tag{5.42}$$

The associated boundary condition is accordingly

$$\begin{aligned}& k_{mtsm}\, A_{surf}\, (C_{sm}(t) - C_{sma}(t) + k_{mtsms}\, A_{su}\, (x_{ssm}(t)\, C_s(t) - x_{ssma}(t)\, C_{sa}(t)) \\ & \quad + k_{mtsmv}\, A_{surf}\, (y_{vsm}(t)\, C_v(t) - y_{vsma}(t)\, C_{va}(t)) \\ & = - D_{smz}\, \partial C_{sm}(t)/\partial z - D_{sz} \partial(x_{ssm}(t)\, C_s(t))/\partial z - D_{vz}\, \partial(y_{vsm}(t)\, C_v(t)//\partial z) \\ & + U_{smz}\, C_{sm}(t) + U_{sz}\, x_{ssm}(t)\, C_s(t) + U_{vz}\, y_{vsm}(t)\, C_v(t))\end{aligned} \tag{5.42a}$$

The net formation of secondary metabolites in the top layer of the wallboard equals the radial diffusion of secondary metabolites in the wallboard, plus the convective transport of secondary metabolites radially in the wallboard, plus the accumulation of secondary metabolites in the circular segment of radial width Δr and thickness Δz, plus the diffusive transport of secondary metabolites in the z dimension, plus the convective transport of secondary metabolites in the z dimension. It is assumed that the secondary metabolites contained in the spores and vegetative cells diffuse or migrate convectively at the same rate as the spores and vegetative cells, respectively.

The secondary metabolites consist of those freely mobile in the top layer of the wallboard, $C_{sm}(t)$; those carried in the top layer or backside paper layer of the wall board by spore particles, $x_{ssm}(t)\, C_s(t)$; and those carried in the top or backside layer of the wallboard by the particles consisting of vegetative cells, $y_{vsm}(t)\, C_v(t)$. The term

$x_{ssm}(t)$ is the fractional content of secondary metabolites in the spore particles, and the term $y_{vsm}(t)$ is the fractional content of secondary metabolites in the vegetative particles.

The boundary condition relates the net transport of secondary metabolites in the air space to the diffusion and convection of secondary metabolites in the z direction into the paper surface.

5.16.3 Secondary Metabolite Balances for the Air Space Adjacent to the Backside Layer of the Wallboard

Similarly, for the backside layer region of the wallboard, dealing with the air space adjacent to the backside layer

$$\begin{aligned} &Q_{maisub}\,(C_{smaisub}\,(t) + x_{maisub}(t)\;C_{saisub}(t) + y_{vmaisub}(t)\;C_{vaisub}\,(t)) \\ &\quad + k_{mtsmsub}\,A_{surfsub}\,(C_{smsub}\,(t) - C_{smasub}\,(t)) + k_{mtsmssub}\,A_{surfsub}\,(x_{ssmsub}(t)\;C_{ssub}\,(t) \\ &\quad - x_{ssmasub}(t)\;C_{sasub}\,(t)) + k_{mtsmvsub}\,A_{surfsub}\,(y_{vsmsub}(t)\;C_{vsub}\,(t) \\ &\quad - y_{vsmasub}(t)\;C_{vasub}\,(t)) = Q_{masub}\,(C_{smasub}\,(t) + x_{masub}(t)\;C_{sasub}\,(t) \\ &\quad + y_{vmasub}(t)\;C_{vasub}\,(t)) + V_{sub}\partial(C_{smsub}\,(t) + x_{ssmsub}(t)\;C_{ssub}\,(t) \\ &\quad + y_{vsmsub}(t)\;C_{vsub}\,(t))/\partial t + Q_{masub}\,(C_{smasub}\,(t) + x_{masub}(t)\;C_{sasub}\,(t) \\ &\quad + y_{vmasub}(t)\;C_{vasub}\,(t)) \end{aligned} \tag{5.43}$$

5.16.4 Secondary Metabolite Balances for the Gypsum Sublayer of the Wallboard

For the gypsum sublayer of the wallboard, the mass balance follows.

$$\begin{aligned} \Omega_{3smsub} &= -1/r\,\partial/\partial r\,(r\,D_{smrsub}\,\partial C_{smsub}\,(t)/\partial r + r\,D_{smrsub}\,\partial(x_{ssmsub}(t)\;C_{ssub}\,(t)\,/\partial r \\ &\quad + r\,D_{smrsub}\,\partial(y_{vsmsub}(t)\;C_{vsub}\,(t)/\partial r) - 1/r\,\partial/\partial r(r\,U_{smrsub}\,(C_{smsub}\,(t) \\ &\quad + r\,U_{srsub}\,x_{ssmsub}(t)\;C_{ssub}\,(t) + r\,U_{vrsub}\,y_{vsmsub}(t)\;C_{vsub}\,(t)) + \partial((C_{smsub}\,(t) \\ &\quad + x_{ssmsub}(t)\;C_{ssub}\,(t) + y_{vsmsub}(t)\;C_{vsub}(t))\,/\partial t + \partial/\partial z(D_{smzsub}\,\partial C_{smsub}\,(t)/\partial z \\ &\quad + D_{srsub}\,\partial(x_{ssmsub}(t)\;C_{ssub}\,(t)/\partial z + D_{vzzsub}\,\partial(y_{vsmsub}(t)\;C_{vsub}\,(t)/\partial z) \\ &\quad + \partial(U_{smzsub}\,C_{smsub}\,(t))/\partial z + \partial(U_{szsub}\,x_{ssmsub}(t)\;C_{ssub}\,(t))/\partial z \\ &\quad + \partial(U_{vzsub}\,y_{vsmsub}(t)\;C_{vsib}(t))/\partial z \end{aligned} \tag{5.44}$$

5.16.5 Equation at the Interface between Top or Backside Paper Layers and the Gypsum Sublayer for the Secondary Metabolite Balances

At the interface between top or backside paper layers and the gypsum sublayer for the secondary metabolite balances

$$-D_{smz}\,\partial/\partial z\,C_{sm}(t) - D_{sz}\,\partial/\partial z\,(x_{sm}(t)\,C_s(t)) - D_{vz}\,\partial/\partial z\,(y_{vsm}(t)\,C_v(t))$$
$$+\,\partial/\partial z\,(U_{smz}\,C_{sm}(t) + U_{sz}\,x_{ssm}(t)\,C_s(t) + U_{vz}\,y_{vsm}(t)\,C_v(t))$$
$$= -D_{smzsub}\,\partial/\partial z\,(C_{smsub}(t) - D_{szsub}\,\partial/\partial z\,(x_{ssmsub}(t)\,C_{ssub}(t))$$
$$-\,D_{vzsub}\,\partial/\partial z\,(y_{vsmsub}(t)\,C_{vsub}(t)) + \partial/\partial z\,(U_{smzsub}\,C_{smsub}(t)$$
$$+\,U_{szsub}\,x_{ssmsub}(t)\,C_s(t) + U_{vzsub}\,y_{vsmsub}(t)\,C_v(t)) \qquad (5.45)$$

where $z = z_p$ and $z = z_{p+1}$.

5.16.6 Secondary Metabolite Formation in the Paper Layer

The reaction terms Ω_{3sm} and Ω_{3smsub} are dependent on the metabolism involving spores and vegetative cells in regard to propensity to form secondary metabolites. It should depend on vegetative growth since release of secondary metabolites may be a consequence of such growth due to the need to release waste products. Secondary metabolites should also form as spores convert to vegetative growth and also when vegetative growth converts to spores. When either spores or vegetative cells are destroyed, we might expect secondary metabolite formation as a form of death release and protective response mechanism (at least hypothetically). We have to assume that the metabolites that form may differ depending on cause. However, we will assume one secondary metabolite formed as a function of different rates tied to spore and vegetative cell rates.

Therefore,

$$\Omega_{3sm} = k_{asm}\,(\mu_s\,C_s) + k_{bsm}\,(\mu_v^1\,C_v) + k_{csm}\,(\mu_{sd}\,C_s) + k_{dsm}\,(\mu_v\,C_v) + k_{esm}\,(\mu_v^1\,C_s)$$
$$+\,k_{fsm}\,(\mu_{vd}\,C_v) - \mu_d\,((C_{sm}(t) + x_{ssm}(t)\,C_s(t) + y_{vsm}(t)\,C_v(t)) \qquad (5.46)$$

5.16.7 Secondary Metabolite Formation in the Gypsum Sublayer

Similarly, for the gypsum sublayer,

$$\Omega_{3smb} = k_{asmb}\,(\mu_{sb}\,C_{sb}) + k_{bsmb}\,(\mu^1{}_{vb}\,C_{vb}) + k_{csmb}\,(\mu_{sdb}\,C_{sb}) + k_{dsmb}\,(\mu_{vb}\,C_{vb})$$
$$+\,\mathrm{k}_{esmb}\,(\mu^1_{vb}\,C_{sb}) + k_{fsmb}\,(\mu_{vdb}\,C_{vb}) - \mu_{dsub}\,(C_{smsub}(t)$$
$$+\,x_{ssmsub}(t)\,C_{ssub}(t) + y_{vsmsub}(t)\,\mathrm{C}_{vsub}(t)) \qquad (5.47)$$

Therefore, the assumption is that all reactions contribute to the reaction formation of secondary metabolites but at different rates. Again, there may be a set of different secondary metabolites formed yielding vector matrices perhaps. Also the reaction rate may be more complex than first order and can depend on other variables as shown later.

Note that terms are included for potential destruction of the secondary metabolite by natural or imposed mechanisms.

5.16.8 Kinetics of Secondary Metabolite Formation

For any k_{ism} or k_{ismb}

$$k_{ism} = (k_{ism0})/(1 + k_{ism1}/N_{ism1} + k_{ism2}/N_{ism2} + k_{iism1}\,I_{iism1} + \ldots) \quad (5.48a)$$

$$k_{ismb} = (k_{ismb0})/(1+ k_{ismb1}/N_{ismb1} + k_{ismb2}/N_{ismb2} + k_{iismb1}\,I_{iismb1} + \ldots) \quad (5.48b)$$

There may be specific factors that limit secondary metabolite formation besides those affecting spores and vegetative cells.

5.16.9 Kinetics of Secondary Metabolite Destruction

$$\mu_d = (k_{dism})/(1 + k_{dism1}/N_{dism1} + k_{dism2}/N_{dism2} + k_{dism3}\,I_{dism3} + \ldots) \quad (5.49a)$$

$$\mu_{dsub} = (k_{dismb})/(1 + k_{dismb1}/N_{dismb1} + k_{dismb2}/N_{dismb2} + k_{dismb3}\,I_{dismb3} + \ldots) \quad (5.49b)$$

There may be specific factors that limit or affect secondary metabolite destruction besides those affecting spores and vegetative cells.

μ is rate coefficient, k is a rate constant, E is enzyme concentration, N is an essential nutrient, and I is a key inhibitor or antagonist for the secondary metabolite.

5.17 DEPLETION OF WALLBOARD FOR MOLD NUTRITION AND SECONDARY METABOLITE FORMATION

We need to account for the finite amount of nutrition available for fungal transformation and growth and product formation in the wallboard since the nutrition will be depleted as the mold forms and yields products. This is based on reactions that occur in the system as defined previously. From before, respectively, Equations 5.1, 5.2, and 5.46:

$$\Omega_{1\ \text{spore transformation}} = \mu_s\, C_s - \mu^1_v\, C_v + \mu_{sd}\, C_s$$

$$\Omega_{2\ \text{vegetative transformation \& growth}} = \mu_v\, C_v - \mu^1_v\, C_s + \mu_{vd}\, C_v$$

$$\begin{aligned}\Omega_{3sm} = {} & k_{asm}\,(\mu_s\, C_s) + k_{bsm}\,(\mu^1_v\, C_v) + k_{csm}\,(\mu_{sd}\, C_s) \\ & + k_{dsm}\,(\mu_v\, C_v) + k_{esm}\,(\mu^1_v\, C_s) + k_{fsm}\,(\mu_{vd}\, C_v) - \mu_d\,((C_{sm}\,(t) \\ & + x_{ssm}(t)\; C_s\,(t) + y_{vsm}(t)\; C_v\,(t))\end{aligned}$$

The following definitions are made:

W_{po} = Initial dry weight of the outer paper layer at the forward or back surface of the wallboard

W_{subo} = Initial dry weight of the gypsum sublayer of the wallboard

For any nutrient quantity, Ni is the system (where Ni is defined as the same particular nutrient for spores changing to vegetative cells, for vegetative cells propagating, or for vegetative cells transforming back to spores).

N_{io} and N_i are the amount of specific nutrient i at time zero and time t, respectively, in the wallboard. The nutrient may be, for example, carbohydrate in cellulose.

v_{io} and v_i are the weight fractions of the nutrient i in the wallboard at time zero and time t, respectively

Therefore,

$$v_{io} W_{po} = N_{ios} \tag{5.49c}$$

$$v_i W_{pi} = N_i \tag{5.49d}$$

At time t,

$$N_i = N_{isvs} + N_{ivs} + N_{ivss} \tag{5.49e}$$

$$v_{isubo} W_{subo} = N_{isubo} \tag{5.49f}$$

$$v_{ib} W_{subi} = N_{isub} \tag{5.49g}$$

At time t,

$$N_{isub} = N_{isvsub} + N_{ivsub} + N_{ivssub} \tag{5.49h}$$

Each of these apply respectively to the nutrient i when spores are being converted to vegetative cells, where vegetative cells are propagating, and to where vegetative cells are converting back to spores. These nutrients may theoretically at least or for completeness, be present in the paper layer or in the gypsum sublayer. In the following equation, j represents the surface layer as s, or the gypsum sublayer as sub.

This equation states that the amount of nutrient i initially in a surface layer or in a sublayer equates that nutrient i remaining at time t from conversion of spores to vegetative cells, N_{isvj}, that remaining at time t from that supplied for vegetative propagation, N_{ivj}, that remaining at time, N_{ivsj}, used for conversion of vegetative cells to spores plus terms for the use of nutrients to form secondary metabolite products. The three remaining terms, respectively, are for the total reactions involved for spores using nutrient i, the total reactions involving nutrient i for use in vegetative cells, and the third term for reactions involving secondary metabolite formation. The term γ_{ij} is the stoichiometric coefficient for the amount nutrient involved per unit of indicated reaction.

$$N_{ioj} = N_{isvj} + N_{ivj} + N_{ivsj} + \gamma_{ij} (\Omega_{1j} + \mu^1{}_{vj} C_{vj}) + \gamma_{ij} (\Omega_{2j} + \mu_{vj} C_{vj}) + \gamma_{ij} (\Omega_{3smj} + \mu_{dj} (C_{smj} + x_{smj} C_{sj} + y_{smj} C_{vj}))$$

This is Equation 5.49i for a surface layer,

$$N_{ios} = N_{isvs} + N_{ivs} + N_{ivss} + \gamma_{is} (\Omega_1 + \mu^1{}_v C_v) + \gamma_{is} (\Omega_2 + \mu_v C_v) + \gamma_{is} (\Omega_{3sm} + \mu_d (C_{sm} + x_{sm} C_s + y_{sm} C_v)) \tag{5.49i}$$

and Equation 5.49j where j is sub, that is,

$$N_{iosub} = N_{isvsub} + N_{ivsub} + N_{ivssub} + \gamma_{isub} (\Omega_{1sub} + \mu^1{}_{vsub} C_{vsub}) + \gamma_{isub} (\Omega_{2sub} + \mu_{vsub} C_{vsub}) + \gamma_{isub} (\Omega_{3smsub} + \mu_{dsub} (C_{smsub} + x_{smsub} C_{ssub} + y_{smsub} C_{vsub})) \tag{5.49j}$$

These equations allow one to track nutrient usage and wallboard weight depletion by time as nutrients are used up. These two equations are specific for nutrients in the wallboard.

Certain nutrients such as oxygen are limited by supply from the atmosphere, for example, so different relationships then apply.

5.18 SCENARIOS FOR TESTING OF EQUATIONS AND IN SILICO ANALYSIS

1. The area is initially quite dry with mold particles, perhaps a mix of spores and vegetative cells, present in the air space. Moisture is introduced into the air space through airflow.
2. A pipe or valve suddenly starts leaking unbeknown to building occupants. This water intrusion may begin to wet and eventually saturate the wallboard. Mold then begins to propagate and cause damage.
3. Problems in 1 and 2 are discovered quickly and drying is started.
4. Problems in 1 and 2 are undiscovered causing more extensive mold growth before discovery and remediation.
5. Decisions on removal of wallboard and initiation of drying are made based on given calculated degrees of contamination. Drying involves use of heater and fans. The degree and timing of drying and effectiveness is discovered.
6. Degrees of mold contamination are assessed by the model and potential impact on mold particle formation and concentration of spores, vegetative cells, and secondary metabolites are made.
7. Reaction mechanisms and transport may be varied many ways to conduct simulations. For example, convection in the wallboard may be limited and diffusive processes may predominate.
8. The gypsum sublayer may be so void of nutrients that although spores penetrate into the sublayer, they do not convert to vegetative cells there, or if they do they revert back or die. Therefore, the sublayer may simply be a wet but inert layer. This can alter the rate of mold propagation and the rate of production of mold secondary metabolites.

5.18.1 Initial Case Scenario

Consider an air space receiving moderate airflow carrying spores and humid (not saturated) air. The space has been initially dry and so is the wallboard in the space. At time zero, the exposure commences. Spores traverse in, some settle to the wallboard and contaminate it. The mold begins to propagate. Assume the infestation is only on one side of the wallboard and not the backside. The temperature is moderate, 25°C. The temperature is then raised to 50°C as fans are installed to dry the space. Assume this continues for a defined period and then ceases for a time period from zero to t_{final} of 30 days during the humid air exposure and then to 10 days for dry air exposure. Airflow during the drying phase is considerately increased and warmed.

Mathematically describe the events. The first operative equation is the spores carried in with the air. From Equation 5.21

$$Q_{si}\, C_{si}(t) + K_{ms}\, A_{surf}\,(C_{ss}(t) - C_{sa}(t)) = Q_{si}C_{sa}(t) + V\,\partial C_{sa}/\partial t$$

$$Q_{si} = 10 \text{ liters/min } (.01 \text{ m}^3\text{/min}) \text{ for 30 days}$$

Air flow circulation is then increased to 1 m^3/min for 10 days.

$$C_{si}(t) = 1 \text{ g/m}^3 \text{ dry weight (the spore inlet concentration)}$$

During the drying phase, air is introduced, considerably less in spore concentration at .001 g/m^3. The latter figure is also the initial spore concentration in the air space before the infestation occurs.

5.18.2 Surface Spore Infestation Assumptions

Assume the surface is 1 m^2. The net transport rate for spores is assumed to be 0.002 cm/sec or 0.0002 m/sec at the initial low air rate. This is assumed based on a diffusion rate of .0001 cm^2/sec over a diffusion barrier of 500 μ. When the air rate is increased during the drying phase, the diffusion rate is assumed to be .0005 cm^2/sec and the diffusion barrier at the air–wallboard surface is assumed to be then 100 μ, yielding a mass transfer coefficient of 0.05 cm/sec, or 0.0005 m/sec.

The initial concentration of spores at the surface, $C_{ss}(t)$, is .001 g/m^2. This assumes that the surface concentration initially of spores per unit area is the same as that per unit volume of air at the low value. The adsorption, Equation 5.21a, is assumed to follow its simplest form for this example. If the surface tends to reject spores or bind them more readily, this figure may vary. It will certainly possibly vary by material of construction and conditions. Wet wallboard may bind spores more readily, for example.

5.18.3 Volumetric and Surface Assumptions for the Case

The exit spore concentration should be obtained by mass balance, depending on degree of change due to surface binding or release. Similarly the accumulation of spores is obtained by calculations that are interactive between the air space and the wallboard surface. Assume the amount of volumetric space is numerically equal to the amount of nonfloor surface exposed (so an 8 ft by 8 ft by 4 ft space or 224 sq ft will have a volume of 256 cubic feet). Therefore, since the basic surface is 1 m^2, the basic volume (V) is 1 m^3.

5.18.4 Mass Balance Assumptions for Vegetative Cells in the Air Space

Similarly, for the mass balance for vegetative cells in the air space (Equation 5.22)

$$Q_{vi}\, C_{vi}(t) + K_{mv}\, A_{surf}\,(C_{vs}(t) - C_{va}(t)) = Q_{vi}C_{va}(t) + V\partial C_{va}/\partial t$$

For this case, we assume that the number of vegetative cells entering, $C_{vi}(t)$, is zero, perhaps postulating that the vegetative cells are not airborne and that only spores are airborne. However, the transport rate of vegetative cells is proposed to be the same as for spores as given earlier for this case.

5.18.5 Case Assumptions for Spore Mass Balance Infestation at the Surface

The mass balance for spores at the wall surface layer is given from before by Equations 5.23 and 5.23a.

Assume for this initial case that multiple spores land uniformly in the 1 m square piece of wallboard. The radial diffusion of spores in wet wallboard paper is assumed to be 10^{-7} cm^2/sec. The initial concentration of spores at the surface before introduction of moist air containing spores is as given, 0.001 g/m^2. Assume that radial convection is nil so that the convective velocity in the radial direction U_{sr} is zero. Assume that the diffusion into the paper and gypsum subsurface is also at 10^{-7} cm^2/sec. Again, convective transport of spores in the z direction is assumed to be zero, so that U_{sz} is zero. The boundary condition for large r should be that C_s approaches the initial condition of 0.001 g/m^2. For example, this can be within 1% of 0.001 g/m^2.

This assumes that radial growth centers do not interact. If spores overlap and transform to vegetative cells that overlap, as a fungal mat will overlap, then their growth should be naturally limited by available nutrition. They may introduce an entity or remediation that controls growth. This way the number of growth centers follows that population concentration. The amount of biomass formed of spores or vegetative cells can be obtained by integrating over an average growth center and then multiplying that figure by a calculation of the amount of growth center present at any point in time.

5.18.6 Spore Loci Assumptions at the Wallboard Surface

The number of spores that have alighted at .0.001g/m^2 is assumed to be 10^6 spores at 10^{-9} grams/spore onto 1 m^2 as the initial condition. If 10^9 spores alight during a condition of contamination onto the 1 m^2, that is, 10^9 spores/m^2. Therefore, every square centimeter will hold about 10^5 spores (and every square millimeter will hold about 10^3 spores) in the state where contamination occurs. If each spore is about 30 μ in size or 0.03 mm, then that is 0.0009 mm^2 projected area per spore.

One spore occupies 0.0009 mm^2 on the average. The total surface occupied as it migrates and transforms and propagates can be calculated by computing migration distances. At any point, we can divide the total area of 1 m^2 by the area ultimately resulting for one spore and its transformants. That figure allows us to project total events for the sample surface of 1 m^2 of wallboard.

At 1000 spores per mm^2, 0.9 mm^2 are occupied per square millimeter of wallboard surface since there are 10^6 mm^2 of surface per square meter of wallboard. At this condition the spores are almost on top of one another and will transform to vegetative growth and propagate without moving.

5.18.7 Moisture Assumptions for the Case

It may be that the moisture alone in the air may be sufficient to cause growth. That needs to be checked as condition of moisture-laden versus nonmoisture-laden air. At 25°C, from Equation 5.40a, the moisture content of the air at saturation is 0.0195kg/kg dry air at 25°C. At 10% saturation, the content is 0.00195 kg/kg dry air. At 90% of saturation, the moisture content is 0.01755.

Transport of water to the wallboard will depend on the equilibrium moisture content of the wallboard as well as the moisture content of the air. In this example, the moisture content is controlled by the moisture in the incoming air. Therefore, from Equation 5.31,

$$Q_a\, C_{wi}(t) + K_w\, A_s(C_w - C_{wa}(t)\, \beta) = Q_a C_{wa}(t) + V\partial C_{wa}/\partial t$$

The airflow in two states is as specified. The mass transfer coefficient for water in the conditions of contamination and then drying is assumed to be as follows.

5.18.8 Diffusion and Transport Assumptions for Spores and Water for the Case

The net transport rate for spores is assumed to be 0.002 cm/sec or 0.00002 m/sec at the initial low air rate. This is assumed based assuming a diffusion rate of .0001 cm^2/sec over a diffusion barrier of 500 μ. When the air rate is increased during the drying phase, the diffusion rate is assumed to be .0005cm^2/sec and the diffusion barrier at the air–wallboard surface is assumed to bc then 100 μ, yielding a mass transfer coefficient of 0.05 cm/sec or 0.0005 m/sec.

Assume for this calculation that the equilibrium relationship between water content in the wall (kg water/kg of wall board) is 0.1 when the air is 100% saturated at the 25°C or at 0.0195 kg water/kg dry air. Therefore, β is equal to 0.1/.0195 or 5 kg water/kg dry wallboard/kg water/kg air. It is assumed that this proportionality holds even at different temperatures. The proportionality also holds when relating the capacity of dry wall to absorb water versus that of air. Therefore, Equation 5.40a also holds for drywall, except the proportionality is assumed to apply. When equilibrium is reached, net water transport between air and wallboard will not occur.

5.18.9 Assumptions for Mass Transfer for Vegetative Cells at the Surface for the Example Case

The mass balance for vegetative cells at the surface is given from before by Equations 5.24 and 5.24a.

The mass transfer coefficient for vegetative cells to the surface is assumed to be as given earlier for spores for this example. Assume for this initial case that multiple vegetative cells land uniformly in the one meter square piece of wallboard. The radial diffusion of vegetative cells in wet wallboard paper is assumed to be 10^{-8} cm^2/sec. The initial concentration of vegetative cells at the surface before introduction of moist air containing spores and vegetative cells is as given, 0.001 g/m^2. Assume that radial convection is nil so that the radial convection velocity for vegative cells U_{vr} is zero. The radial motion is assumed to be diffusive in nature. Diffusion is based on active growth and spread. Assume that the diffusion into the paper and gypsum subsurface is also at 10^{-8} cm^2/sec. Again, convective transport of vegetative cells in the z direction is assumed to be zero, so that U_{vz} is zero. The boundary condition for large r, should be that C_v approaches the initial condition of 0.001 g/m^2. For example, this can be within 1% of .0.001 g/m^2.

5.18.10 Interaction of Radial Growth Centers at the Surface

This assumes that radial growth centers do not interact. If they do, the flux boundary condition applies (where one growth center begins to overlap another) and will be limited by opposing concentration gradients. The radial span can be inferred by dividing the number of cells into a square meter of surface and size of a cell as opposed to say a square millimeter of space. Since the cells propagate and spread, they need room to grow. At low concentrations of cells, there will be ample room so the zero boundary condition is appropriate. At larger concentrations as noted the fungal mats may overlap limited by available nutrition in the wallboard, or in the case of oxygen, in the surrounding air space. As noted, the cells will compete for nutrients locally and that may prevent them piling up on one another. The mass balances and calculations should point this out.

5.18.11 Management of Loci for Vegetative Cells at the Surface for the Case Scenario

As noted earlier in section 5.18.6, the number of vegetative cells alighting at 0.001g/m^2 is assumed to be 10^6 cells at equal to10^{-9} grams/cell onto 1 m^2 as the initial condition. If 10^9 cells alight during a condition of contamination onto the 1 m^2, that is equal to 10^9 cells/ m^2. Therefore, every square centimeter will hold about 10^5 cells (and every square millimeter will hold about 10^3 cells) in the state where contamination occurs. If each cell (not changed in weight or size significantly from that of a spore in this example) is about 30 μ in size or 0.03 mm, the projected area is 0.0009 mm^2. At 1000 cells per mm, 0.9 mm^2 occupied per square millimeter of wallboard surface since there are 10^6 mm^2 of surface per square meter of wallboard. At this condition the cells are almost on top of one another and will propagate without moving. If they are stressed or starved at any locale, they will revert to spores again.

5.18.12 Boundary Layer Assumptions between Paper Surface and Gypsum Sublayer

At the position, between the peripheral layers and the gypsum sublayer, a flux boundary condition results (the z direction). For this case, diffusion and reaction can also occur in the sublayer depending on nutrient availability and water penetration. The layer at the backside outer paper surface is exposed to air flowing slowly but without excessive water in it. It may pick up water as well as spores and cells at the lower rate and thereby act to dry the wallboard to a degree.

Assume water vapor diffusivity (D_{wr} and D_{wz}) in the radial and z directions to be 1e-05 cm^2/sec in the paper layers and 4e-04 cm^2/sec in the gypsum sublayer.

5.18.13 Reaction Terms for Spores and Vegetative Cells for the Sample Case Analysis

As covered in Equations 5.1, 5.2, and 5.3, the reaction terms are given as follows:

$$\Omega_{1\ \text{spore transformation}} = \mu_s C_s - \mu^1{}_v C_v + \mu_{sd} C_s$$

$\mu_s = 5.5$ days^{-1}, $\mu^1{}_v = 5.5$ days^{-1} and $\mu_{sd} = 0$, meaning spores and vegetative cells convert to each other at the same rate per unit respective concentration and spores do not die as spores are difficult to kill.

$$\Omega_{2\ \text{vegetative transformation \& growth}} = -\mu_v C_v - \mu_s C_s + \mu^1{}_v C_v + \mu_{vd} C_v$$

$\mu_v = 0.73$ days^{-1}, $\mu_s = 5.5$ days^{-1}, $\mu^1{}_v = 5.5$ days^{-1} and $\mu_{vd} = 0$, meaning vegetative growth is much slower than spore transformation, and spores and vegetative cells convert to each other at the same rate per unit respective concentration. Vegetative cells are easier to kill. However, in this example, assume a kill mechanism is not imposed.

5.18.14 Reaction Parameter Assumption for Secondary Metabolites for the Case Scenario

The reaction rate mechanism for secondary metabolite formation (expressed as an addition on the left-hand side of mass balance equations) is (for the surface paper; from Equation 5.46):

$$\Omega_{3sm} = k_{asm} (\mu_s C_s) + k_{bsm} (\mu^1{}_v C_v) + k_{csm} (\mu_{sd} C_s) + k_{dsm} (\mu_v C_v) + k_{esm} (\mu_s C_s)$$
$$+ k_{fsm} (\mu^1{}_v C_v) + k_{gsm} (\mu_{vd} C_v) - \mu_d ((C_{sm}(t) + x_{ssm}(t) C_s(t) + y_{vsm}(t) C_v(t))$$

Each reaction is the system can plausibly be associated with secondary metabolite formation. Therefore, for the gypsum subsurface (from Equation 5.47),

$$\Omega_{3smb} = k_{asmb} (\mu_{sb} C_{sb}) + k_{bsmb} (\mu^1{}_{vb} C_{vb}) + k_{csmb} (\mu_{sdb} C_{sb}) + k_{dsmb} (\mu_{vb} C_{vb})$$
$$+ k_{esmb} (\mu^1{}_{vb} C_{sb}) + k_{fsmb} (\mu^1{}_{vb} C_{vb}) + k_{gsmb} (\mu_{vdb} C_{vb}) - \mu_d (C_{smsub}(t)$$
$$+ x_{ssmsub}(t) C_{ssub}(t) + y_{vsmsub}(t) C_{vsub}(t))$$

Secondary metabolite concentrations are typically small relative to spore or cell concentrations. Therefore, let each of the k terms be 0.001.

5.18.15 Parameters for Control of Germination and Sporulation in the Case Scenario

For germination in Equation 5.4, for the first case, we set all k parameters to zero assuming that the water content parameter is controlling.

Therefore,

$$\mu_s = (\mu_{s0})/(1 + k_{sw}/C_{sw1})$$

The organism adjusts for germination and for sporulation. Under germination when water content (C_{sw1}) is elevated to 0.1 weight of water per weight of wallboard paper (reflects exposure to saturated water content of 0.0195 g water/g dry air at 25°C), k_{sw} is 0.01 so that the rate of germination of spores (5.5/day) is about 90% of the maximum rate germination (μ_{s0}). Under conditions where water is low, germination should be inhibited by organism changes to metabolism so that at a moisture content in the wallboard of 0.01 grams water/gram of wallboard, then k_{sw} is 0.1. This reflects exposure to air that is 10% saturated at 25°C (0.00195 grams of water/gram dry air, amount of water in the wallboard/unit weight then being five times that in the air, grams water per grams of dry air, or β of 5).

For vegetative reversion to spores, conditions change and we wish to induce sporulation. The relationship for rate of sporulation is given in Equation 5.8. We set all k parameters to zero, except for the one involving water, since we are making the assumption for this case that only control of the rate of sporulation in response to water content is critical. Therefore,

$$\mu^1_v = \mu^1_{vs0}/(1 + k_w/C_w)$$

For sporulation to occur, water content must be low, for example, exposed to air that is 10% saturated with water. This is reflected in a moisture content in the wallboard paper of 0.01 grams water/gram of wallboard. For this rate to be proximate to the maximum rate, $\mu^1_v \sim \mu^1_{vs0}$, then k_w is 0.001.

When the moisture content rises, the organism metabolism must be such as to inhibit sporulation. So if the moisture content becomes 0.1 grams water/gram wallboard, then k_w is assumed to be increased to 1.

In the sublayer, where spores can germinate and vegetative cells can sporulate, we apply similar assumptions to parameters as applied for the paper layer. Therefore, in Equation 5.13, all k parameters are set to zero implying only water is critical for germination. Similarly in Equation 5.19, all k parameters are set to zero except for the one involving water.

Therefore, in the equation for germination in the sublayer, applying Equation 5.13 in this manner, $\mu_{sb} = (\mu_{sb0})/(1 + k_{sbw}/C_{sbw1})$. At high water content, where $C_{sbw1} = 0.1$, the organism is assumed to set metabolic conditions so that $k_{sbw} = 0.01$ so that germination can proceed relatively uninhibited. In the case where water content is low, then the organism will change matters to inhibit germination. Therefore, when $C_{sbw1} = 0.01$, $k_{sbw} = 0.1$ so that germination will be inhibited.

Accordingly, when water content is low, sporulation in the sublayer will be favored and germination inhibited. Equation 5.19, $\mu^1_{vb} = \mu^1_{vsb0}/(1 + k_{wb}/C_{wb})$, covers sporulation in the sublayer as indicated for k parameters set to zero except for the one involving water. At low moisture content in the sublayer, at 0.01 grams water/gram gypsum layer, k_{wb} is set at 0.001 so that the rate, μ^1_{vb}, will be proximate to the maximum rate, μ^1_{vsb0}. Conversely, when the water content is high, sporulation should be inhibited. Therefore, at 0.1 grams water/gram gypsum layer, C_{wb}, the value for k_{wb} is set to 1.

5.18.16 Nutritional Factors Affecting Reaction Rates in the Case Scenario

A main nutritional component is a carbon source. The paper layers are of course mostly cellulose and therefore high in carbon content. The gypsum layer is mostly devoid of carbon content. Assume the wallboard that has been set as 1 m² in area is ½ inch thick (1.27 cm). Assume the paper layers to each be 0.1 mm or 0.01 cm thick leaving 1.25 cm thickness for the gypsum sublayer. If the density of gypsum sublayer and paper layers are each 0.6 g/cc dry basis, then the gypsum layer consists of 7,500 grams in 12,500 cc. Assume that the gypsum is 1% carbon source. Therefore, the carbon source content of the gypsum is 75g/12,500 cc or 0.006 g/cc. A paper layer consists of 10,000 sq cm by 0.01 cm or 100 cm³ in volume. Assuming its density is also 0.6 g/cc, then each paper layer weighs 60 g. Assume that the paper layer weight consists of 95% available carbon content. Therefore, the carbon content of the paper is 0.57 g/cc. Assume from other fungal growth data that one gram of fungal biomass contains 0.5 g of carbon content, so that γ is 0.5.

Starting at 1 mg of spores per square meter, assuming these are converted to vegetative cells, by rates assumed for propagative growth, in one week, at an increase of 0.73/day, the result will be 165 mg/square meter. On a volumetric basis, this will be 165 mg/100 cm³ or 1.65 mg/cm³, or 0.00165 g/cc. At a γ of 0.5, this will require 0.000825 g of carbon/cc. Since the carbon content of the paper is 0.57 g/cc, very little of the carbon content has been used. At 1 g/square meter loading, a factor of 1,000 increase results. Therefore the carbon use is now 0.825 g/cc. This exceeds the carbon content in the paper so the propagation will not occur at near the maximal rate and will be inhibited to an extent.

In the gypsum sublayer, the carbon content is 0.006 g/cc. The requirement of 0.000825 g of carbon/cc for the propagation is a substantial fraction of the carbon and inhibition of the reaction due to nutrition limitation is expected. Given this, the equations just covered for water content parameters must be modified for this nutritional limitation for the sample case. Carbon nutrition will be required for germination as well as for sporulation. For the latter, the organism may cannibalize its own stores if the nutrition is not available from the environment.

Consider first germination in Equation 5.4, keeping the water and carbon nutrition parameters while setting the others to zero.

$$\mu_s = (\mu_{s0})/(1 + k_{sw}/C_{sw1} + k_c/N_c)$$

For our example, the carbon nutrition availability in the paper layer is 0.57 g/cc and 0.006 g/cc in the gypsum layer. In the paper layer, set k_c to 50% of the maximum available carbon of 0.57 g/cc or 0.285 g/cc.

Equation 5.13 is for germination in the gypsum sublayer. Examine this equation again, keeping the water and carbon nutritional parameters while setting the other parameters to zero.

$$\mu_{sb} = (\mu_{sb0})/\ (1 + k_{sbw}/C_{sbw1} + k_{csub}/N_{csub})$$

For our example, since the carbon nutrition availability in the sublayer is 0.006 g/cc, set k_{csub} to 0.003 g/cc, again 50% of the maximal value.

Equation 5.8 is for sporulation in the paper layer. Examine this equation again, keeping the water and carbon nutritional parameters while setting the other parameters (for this case scenario) to zero.

$$\mu^1{}_v = \mu^1{}_{vs0}/(1 + k_{vsc}/N_{vsc} + k_w/C_w)$$

Again, set k_{vsc} 50% of the maximum available carbon content of 0.57 g/cc or 0.285 g/cc.

Equation 5.19 is for sporulation in the gypsum sublayer. Examine this equation again, keeping the water and carbon nutritional parameters while setting the other parameters (for this case scenario) to zero.

$$\mu^1{}_{vb} = \mu^1{}_{vsb0}/(1 + k_{vscsub}/N_{vscsub} + k_{wb}/C_{wb})$$

Again, set k_{vscsub} to 50% of the maximum available carbon content in the sublayer or 0.003 g/cc.

In both the paper layer and in the gypsum sublayer, vegetative cells, once formed from spores, will propagate. This propagation will be limited by nutritional content. In this case, we are focusing on carbon source. Other limitations may be apparent that can be covered in other cases depending on analytical data, which is quite sparse. For the surface paper layer, governed by Equation 5.6, set all k's to zero except the one involving carbon source. The parameter, k_{v1}, is set to 50% of the maximal carbon source in the paper or to 0.285 grams carbon/cc of paper.

$$\mu_v = (\mu_{v0})/(1 + k_{vg}/N_{vg})$$

In the gypsum sublayer, propagation is governed by Equation 5.16. Therefore, to focus on carbon source, set all k parameters to zero except for the one for carbon.

$$\mu_{vb} = (\mu_{vb0})/(1 + k_{vbc}/N_{vbc})$$

For the gypsum sublayer, governed by Equation 5.16 as modified, with all k parameters set to zero, except for one for carbon, set k_{vbc} to 50% of the maximal value or 0.003 grams carbon/cc gypsum sublayer.

5.18.17 Carbon Nutritional Content and Access, Case Scenario Assumptions

The mass balance for nutritional content, in this case for carbon, will be used to set the carbon content terms at any point in time during analysis. This is represented by Equation 5.49d. When considering only carbon utilization for biomass formation, only the following terms apply for spore germination and sporulation in the paper layers and gypsum sublayers.

$$N_{ioj} = N_{isvj} + N_{ivj} + N_{ivsj} + \gamma_{ij}\,(\Omega_{1j} + \mu^{1}_{vj}\, C_{vj}) + \gamma_{ij}\,(\Omega_{2j} + \mu_{vj}\, C_{vj})$$
$$+ \gamma_{ij}\,(\Omega_{3smj} + \mu_{dj}\,(C_{smj} + x_{smj}\, C_{sj} + y_{smj}\, C_{vj}\,))$$

This is Equation 5.49d for a surface layer, j being s, and Equation 5.49e for the gypsum sublayer, j being sub. In this case, i represents carbon source. The term on the left is 0.57 g/cc for the paper surface and 0.003 grams/cc for the gypsum sublayer.

Just focusing on carbon and ignoring reaction terms, the total carbon initially equals that remaining at time *t* plus that used to germinate cells, to form spores from germinated cells, and that used to propagate the cells in either the surface or the sublayer. To compute total change in mass or nutrient content, take each spore deposit as the center of a colony, compute changes with time, which include wallboard digested, and multiply by the number of spores that alighted or were present at the beginning of a period.

5.18.18 Enzymes to Release Nutrients: Case Scenario Analysis

It is expected that enzyme activity is restricted to extracellular enzymes released in the vicinity of the organism. The activity such as that to degrade cellulose (e.g., cellulase) is sufficiently active to not be a restricting factor. Imposition of equations dealing with such activity is assumed not necessary for this case scenario. Cellulose conversion to glucose should be rapid. However, access to this sugar by the organism is expected to be limiting due to rate processes in uptake based on available concentration. This can be an area to control and restrict mold function by way of enzyme inhibitors since cellulase inhibitors, for example, have been reported in the literature. It may be that other enzymes, such as laccases and proteases, aid in hydrolysis. Given this, inhibitors of their function may also be relevant.

5.19 TEMPERATURE DEPENDENCE OF REACTIONS

Reactions involving sporulation, conversion between spores and vegetative growth, vegetative growth, and spore or cell destruction will have a temperature dependence. This is important for the nature of the severity of an infestation and deriving means to counter the infestation. For example, a temperature cycle in the presence of an agent may be a means to destroy the microbial entity or counter it. Any reaction in the set of equations that is not composed of other reactions will likely follow an Arrhenius relationship. The typical relationship to be accounted for as a parameter will be of the form:

$$k_{ijkl} \text{ or } \mu_{ijkl} = k_{ijklo} \text{ or } \mu_{ijklo}\ (\exp\,(-E_{ijkl}/RT)$$

where E_{ijkl} is the activation energy.

Available data on temperature dependence of fungal reactions will be used to apply the above relationship in the model.

5.20 THE SYSTEM OF TRANSPORT EQUATIONS

The equations describing the propagation of spores (Equation 5.23), vegetative cells (Equation 5.24), water (Equation 5.32), temperature (Equation 5.37), and secondary metabolites (Equation 5.42) are all special cases of the general transport equation in axisymmetrical coordinates

$$S\frac{\partial \phi}{\partial t}+\frac{1}{r}\frac{\partial}{\partial r}(rU_r\phi)+\frac{\partial}{\partial z}(U_z\phi)=\frac{1}{r}\frac{\partial}{\partial r}\left(rD_r\frac{\partial \phi}{\partial r}\right)+\frac{\partial}{\partial z}\left(D_z\frac{\partial \phi}{\partial z}\right)-\Omega \qquad (5.50)$$

where f represents the quantities being transported. In the case of the transport of spores, f is replaced by C_s. Similarly f is replaced by C_v for transport of vegetative cells, C_w for water transport, and C_{sm} for the transport of secondary metabolites. The equation also applies to the heat transport equation formulated in terms of the temperature, with f representing T.

Note that Equation 5.50 applies to all of the regions of the wallboard, whether paper or the gypsum core. Although the parameters in the equation will vary from one area to the next, the equation remains the same. As described before, it is assumed that the dependent variable will be continuous and that at the interfaces the gradients will vary to assure conservation of mass.

The coefficients U, D, S, and Ω in Equation 5.50 represent the parameters controlling the rates of convection, diffusion, storage, and the source term, respectively. The S term represents the resistance to change in the dependent variable, which is unity for the concentration equations and the heat capacity for the temperature equation. The U term represent convection and is basically a velocity. The velocity through competent wallboard should be practically zero, but if the wallboard is degraded by, for example, moisture, it is possible that the convection term may be significant. The D term represents diffusion and is basically the diffusivity of the material.

The system can be represented as a set of coupled differential equations. The coupling occurs primarily through the source term Ω. In general the set of equations cannot be solved analytically, and a numerical solution procedure is required. A number of different techniques can be used, but here the finite difference technique will be described as it is the most intuitive.

Equation 5.50 in its most general form can be solved numerically. However, it is instructive to consider the extent to which it simplifies in most instances.

Subscripts are used to denote that, in general, the coefficients are anisotropic. For example the rate of diffusion in the radial direction D_r and in the axial direction D_z are different, but in practice it is often assumed that the coefficient of diffusion is isotropic, so that $D_r = D_z = D$.

By contrast, the convection terms are decidedly anisotropic. The coefficients U_r and U_z represent velocities that dominate the direction in which convection takes place. It is often argued that in the geometry under consideration, there is convection through the wallboard in a direction parallel to the z axis. Such convection can be caused by a pressure differential between the front and back of the wallboard. However, to generate convection in the radial direction, a pressure gradient must radiate outward from one location. Such a situation can be generated where a point source occurs near the wallboard. However, it is intuitively clear that given the fact that wallboard is generally large and thin, as a general matter it is much more likely that convection will occur axially through the wallboard from front to back rather than radially inside the wallboard. Therefore in most instances $|U_r| > |U_z|$ and for all practical purposes $U_r \approx 0$. Conservation of mass further requires that if $U_r = 0$, then U_z must be a constant unless factors such as density changes are considered.

In Equation 5.50, the coefficients U, D, C, and Ω can vary spatially and also with time. However, as a practical matter, we expect the diffusive coefficient D to be a function only of z to represent the different materials that make up the wallboard. The D term can therefore be considered piecewise constant, that is it attains one value in the paper and another value in the gypsum core, with a sharp transition between the paper and gypsum core.

If D is, in fact, an isotropic and piecewise constant, U_z constant, and U_r zero, then Equation 5.50 simplifies to

$$S\frac{\partial \phi}{\partial t} + U_z \frac{\partial \phi}{\partial z} = D\left(\frac{1}{r}\frac{\partial}{\partial r}\left(r\frac{\partial \phi}{\partial r}\right) + \frac{\partial^2 \phi}{\partial z^2}\right) - \Omega \tag{5.51}$$

To obtain an analytical solution to Equation 5.50, a simplification such as shown in Equation 5.51 would be crucial to making the equation solvable. However, in applying a numerical solution to the problem, such a simplification is not required or necessarily advantageous. When using a finite difference solution method, all that is required is that the coefficients can be approximated as piecewise constant. In this application, it was argued above that the coefficients are essentially constant in the paper and the gypsum, and therefore the approximation is appropriate.

However, Equation 5.51 is instructive to consider intuitively. If U_z is zero, Equation 5.51 is simply the heat equation, albeit inhomogeneous if Ω is not zero. Therefore the transport is purely driven by diffusion or generation via the Ω term. In competent wallboard, U_z would be essentially zero and therefore this would be the dominant transport mechanism.

Should the wallboard become wet, the competence of the gypsum is often compromised and the wallboard may become crumbly, especially if it is subsequently dried. In such a situation a pressure differential across the wallboard will result in flow across the wallboard. Such pressure differentials could be caused by differential heating, wind, and any number of other common phenomena. It is intuitively clear from Equation 5.51 that the convection term would very quickly become a dominant term. In fact, for $|U_z|$?1 it is intuitively clear that this term would completely dominate the equation and therefore would require that $\frac{\partial \phi}{\partial z} \approx 0$. This basically means that f is the same throughout the thickness of the wallboard, and the concentration would depend on the concentration boundary condition at the upwind surface.

This is what you would expect. If the wallboard looses its competence, the transport and propagation of the mold in either spore or vegetative form would be dominated by convective transport, which is fast and to some extent unpredictable because the convection currents are a factor of ambient winds, temperature gradients, and similar factors that are hard to quantify. However, this would generally indicate catastrophic failure of the wallboard in which case the most prudent course of action would be large-scale removal instead of analysis. We therefore consider the typically more insidious case where the wallboard is largely intact.

5.21 DISCRETE EQUATIONS

Solving Equation 5.50 by analytical methods is only possible for very special cases. In general, numerical methods must be used to solve Equation 5.50. There are numerous ways by which Equation 5.50 can be solved numerically. Each method has advantages and disadvantages, but in the hands of a competent practitioner all the methods can solve Equation 5.50 correctly.

The method presented here is the finite difference method. This method is relatively intuitive in that it uses finite differences to approximate derivatives. For example, $\frac{\partial \phi}{\partial z}$ is approximated using $\frac{\phi_E - \phi_P}{z_E - z_P}$ where the E and P subscripts represent the value of *f* at discrete locations. In essence, the finite difference method does not attempt to solve for *f* everywhere, but rather allows the user to evaluate *f* at an arbitrarily large number of locations.

The finite difference method transforms Equation 5.50 into a set of discrete linearized equations. The set of equations can then be solved using well-known methods. Specifically, this set of equations is well suited to solving on a digital computer. A detailed description of the steps outlined next can be found in many standard texts on the subject.

Figure 5.7 shows a generic, axisymmetric control volume considered by the finite difference equations. The control volume is denoted by the subscript *P*, while the neighboring control volumes in the *r* direction are denoted by subscripts *N* and *S* and in the *z* direction by *E* and *W*. Symmetry is assumed around the axis of rotation, and it is assumed that the angular width of the cell is one radian. The generic cell has a width in the *z* direction of Δz and the distance to the neighboring cells are dz_W and dz_E, respectively. The width of the cell in the *r* direction is Δr and the distance to the neighboring cells are dr_S and dr_N, respectively. The area where this cell joins the neighboring cells are denoted by A_w, A_e, A_n, and A_s and the volume of the cell by *V*. Simple geometry shows that

$$A_e = \frac{1}{2}\left(r_n^2 - r_s^2\right) \tag{5.52}$$

$$A_w = \frac{1}{2}\left(r_n^2 - r_s^2\right) \tag{5.53}$$

$$A_n = r_n \Delta z \tag{5.54}$$

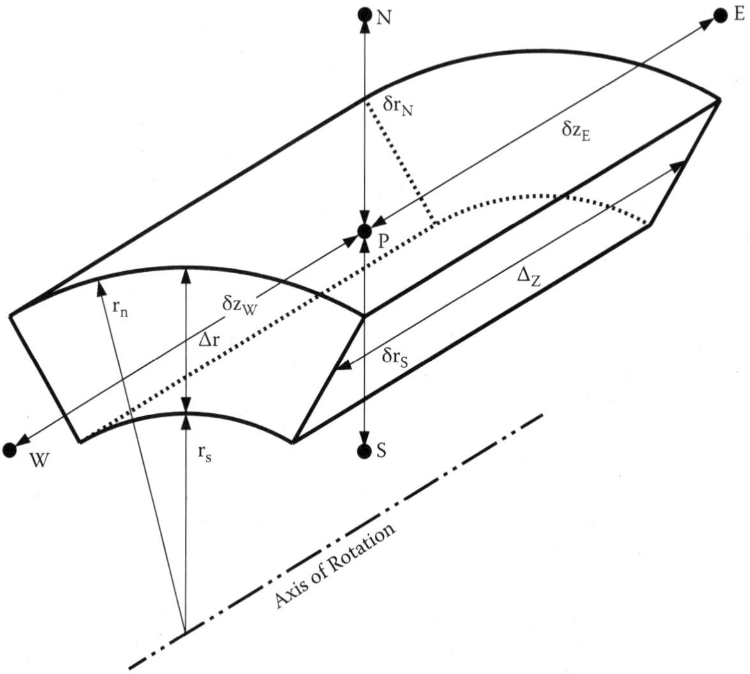

FIGURE 5.7 General cell.

$$A_s = r_s \Delta z \tag{5.55}$$

$$V = \frac{1}{2}\left(r_n^2 - r_s^2\right)\Delta z \tag{5.56}$$

Integrating Equation 5.50 over the control volume and a time step from t_0 to t_1 yields

$$0 = \int_{t_0}^{t_1}\int_{z_w}^{z_e}\int_{r_s}^{r_n} S\frac{\partial\phi}{\partial t} + \frac{1}{r}\frac{\partial}{\partial r}(rU_r\phi) + \frac{\partial}{\partial z}(U_z\phi)$$

$$-\frac{1}{r}\frac{\partial}{\partial r}\left(rD_r\frac{\partial\phi}{\partial r}\right) - \frac{\partial}{\partial z}\left(D_z\frac{\partial\phi}{\partial z}\right) + \Omega\, drdzdt$$

$$= \frac{1}{2}\left(r_n^2 - r_s^2\right)\Delta z S\phi\,|_{t_0}^{t_1}$$

$$+\Delta z\Delta t\left(U_r\phi - D_r\frac{\partial\phi}{\partial r}\right)|_{r_s}^{r_n}$$

$$+\frac{1}{2}\left(r_n^2-r_s^2\right)\Delta t\left(U_z\phi-D_z\frac{\partial\phi}{\partial z}\right)\Big|_{z_w}^{z_e}$$

$$+\frac{1}{2}\left(r_n^2-r_s^2\right)\Delta z\Delta t\Omega \tag{5.57}$$

This form of the equation is called the implicit form, since except for the first term, the remainder of the terms are evaluated at time t_1. The equation also assumes that some of the terms are constant over the cell. In general this is not strictly true, but it is an appropriate approximation for a small enough cell.

Note that Equation 5.57 evaluates $U_z\phi-D_z\frac{\partial\phi}{\partial z}$ and $U_r\phi-D_r\frac{\partial\phi}{\partial r}$ at the edges of the cell. One may attempt to evaluate these terms through simple linear interpolation between the nodes. However, convection makes this a very poor choice. For example, when U_z is large and positive, there is a strong flux from P to E, and therefore the value of f at the boundary f_e will depend strongly by f_P, while f_E will have very little influence on f_e.

To capture this behavior, consider the simplified one-dimensional version of Equation 5.50,

$$U\frac{\partial\phi}{\partial x}-D\frac{\partial^2\phi}{\partial x^2}=0 \tag{5.58}$$

which ignores the transient and source terms. For constant U and D it is readily verified that

$$\phi=A+Be^{\frac{U}{D}x} \tag{5.59}$$

is a solution to Equation 5.58, where A and B are constants. Substituting the areas shown above and the one-dimensional solution Equation 5.59 into Equation 5.57 and finding appropriate values for A and B to interpolate between P and each of the neighboring nodes yields

$$SV\left(\phi_P-\phi_P^0\right)=\frac{A_nU_r}{e^{P_r}-1}(\phi_N-\phi_P)+\frac{A_sU_re^{P_r}}{e^{P_r}-1}(\phi_S-\phi_P)\cdot$$

$$+\frac{A_eU_z}{e^{P_z}-1}(\phi_E-\phi_P)+\frac{A_wU_ze^{P_z}}{e^{P_z}-1}(\phi_W-\phi_P)+V\Omega \tag{5.60}$$

where P_r and P_z are the Peclet number representing the relative contribution of convection and diffusion defined as

$$P_r=\frac{U_r}{D_r/\delta r}, \tag{5.61}$$

$$P_z=\frac{U_z}{D_z/\delta z}. \tag{5.62}$$

Note also that all the values are evaluated at time t_1 except for ϕ_P^0, which represents f_P at time t_0.

Rearranging the terms in Equation 5.60 yields

$$a_P\phi_P = a_E\phi_E + a_W\phi_W + a_N\phi_N + a_S\phi_S + a_0\phi_P^0 + O \tag{5.63}$$

where

$$a_E = A_e \frac{U_z}{e^{P_z} - 1} \tag{5.64}$$

$$a_W = A_w \frac{U_z e^{P_z}}{e^{P_z} - 1} \tag{5.65}$$

$$a_N = A_n \frac{U_r}{e^{P_r} - 1} \tag{5.66}$$

$$a_S = A_s \frac{U_r e^{P_r}}{e^{P_r} - 1} \tag{5.67}$$

$$a_0 = \frac{VS}{\Delta t} \tag{5.68}$$

$$O = V\Omega \tag{5.69}$$

$$a_P = a_E + a_W + a_N + a_S + a_0 \tag{5.70}$$

The importance of convection is readily seen in Equation 5.64 to Equation 5.67. For example, if $U_z >> 1$, then $e^P >> U_z$ and $a_E \approx 0$, while $a_W \approx U_z$.

In the absence of convection, $e^P - 1 = 0$, so that Equation 5.64 to Equation 5.67 degenerate since both the numerator and denominator becomes zero. L'Hôpital's rule can be used to determine the value of these coefficients as $P \to 0$, however it can be readily shown that in the absence of convection, straight-forward linear interpolation can be used in Equation 5.57, which yields

$$a_E = \frac{A_e D_z}{\delta z_E} \tag{5.71}$$

$$a_W = \frac{A_w D_z}{\delta z_W} \tag{5.72}$$

$$a_N = \frac{A_n D_r}{\delta r_N} \tag{5.73}$$

$$a_S = \frac{A_s D_r}{\delta r_S} \tag{5.74}$$

The finite difference formulation in Equation 5.63 in fact yields a system of linear equations that can be solved using standard techniques. When the coefficients in Equation 5.63 and the source term Equation 5.69 are not a function of f, the solution is the solution found by the linear solver. However, when the coefficients in Equation 5.63 or the source term Equation 5.69 are a function of f, Equation 5.63 only represents a linearized version of Equation 5.50. In such an instance it is necessary to use Picard iterations, which iteratively solves the linearized version of the equation, until the solution to the more general nonlinear problem is found.

In this instance, the numerical solution procedure is further complicated by the fact that we do not have a single equation, but five simultaneous equations. Since f represents C_s, C_v, C_w, C_{sm}, and T, it is necessary to solve all five equations simultaneously. This introduces another level of Picard iterations required to simultaneously solve all the equations.

5.22 NUMERICAL RESULTS

Numerical results obviously depend on the specific parameters used in solving the equations. However, numerical experiments are interesting and can lead to insights into the equations.

Figure 5.8 shows the results of one such numerical experiment. The results are shown at three locations. The first location shown as solid lines is at the surface where the spores first landed. The second location shown as dotted lines is at the paper–gypsum interface. The third location shown as dashed lines is in the interior.

Figure 5.8 represents time along the horizontal axis and concentrations along the vertical axis. Spore concentrations are shown in red. Vegetative concentrations are shown in green. The concentration of nutrients are shown in blue.

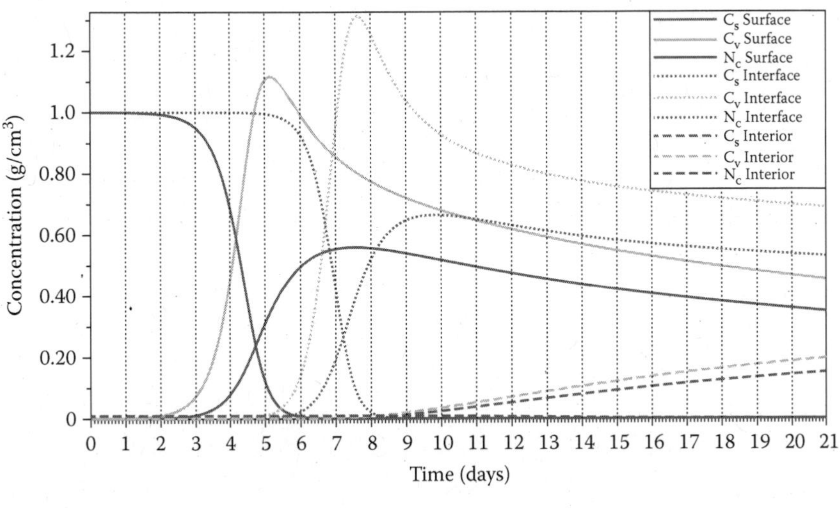

FIGURE 5.8 (See color insert.) Model-predicted concentrations.

At time zero, nutrients are initialized at a concentration of one in the paper layer, and 0.01 in the gypsum layer. The vegetative growth concentrations are initialized to zero. The spore concentrations are initialized to zero everywhere except at the surface, where it is initialized to 0.001 g/cm^3. This represents the initial source of mold.

At the surface location, the spores initially convert to vegetative cells. The vegetative cells then start to grow. Until about four days, the nutrient concentrations remain sufficiently high that vegetative growth is exponential. Between four and six days, vegetative cells continue to grow rapidly but around five days nutrients are essentially exhausted and growth slows and eventually becomes negligible.

Spore concentrations increase exponentially until about five days. The increase in spore concentrations are driven by the conversion from vegetative cells into spores. In the period from about five to eight days, there is a transition to equilibrium, which exists until the end of the simulation. Once the nutrients are exhausted, no further growth occurs, and the mass of vegetative cells and spores remain constant. The ratio between vegetative cells and spores are determined by the rate of conversion between the two forms. From about eight days the surface concentrations decrease due to diffusion away from this location.

At the location at the paper–gypsum interface, the pattern of growth is essentially the same as at the surface. Mold arrives at this location as both spores and vegetative cells. Once the vegetative cell concentration is nonzero, the exponential growth in vegetative cells occurs. This in essence occurs from time zero until about seven days. Due to the abundant nutrients in the paper layer, concentrations at this location actually exceed concentrations at the surface. Spore concentrations also concentrate due to both conversion from vegetative cells and diffusion. As at the surface, concentrations decrease over time due to diffusion into the interior.

In the interior of the gypsum layer, very little nutrients exist. Therefore the period of exponential growth is very limited before nutrients are exhausted. The increase in concentrations at this location is primarily the result of diffusion from the paper layer.

Figure 5.9 shows the sensitivity of the model predictions to the rate of vegetative growth at the surface location where spores are first deposited. Figure 5.9 shows the model predicted concentrations when μ_{v0} is changed from 5.5 to 11 days^{-1}. As expected, the rate of growth doubles, leading to a doubling of the slope on the vegetative concentrations during the initial growth phase. However, nutrients are also depleted at twice the previous rate. Therefore the vegetative cell concentration peaks at approximately 2.5 days, whereas before the peak occurred at about 5 days. The spore concentration is driven by the conversion from vegetative cells and shows a similar acceleration in the growth of concentrations. Once the nutrients are exhausted, the behavior is very similar. Again this is what is to be expected since μ_{v0} is essentially zero at this point.

An interesting observation is that the peak concentration is higher in the case of the higher growth rate. This behavior can be attributed to the fact that the growth is exponential, while diffusion is proportional to the concentration gradient. Therefore the local concentration peak is sharper as rapid growth elevates the concentrations. Then, as the nutrients are exhausted, the steeper concentration gradient leads to more rapid diffusion away from this location. Since the amount

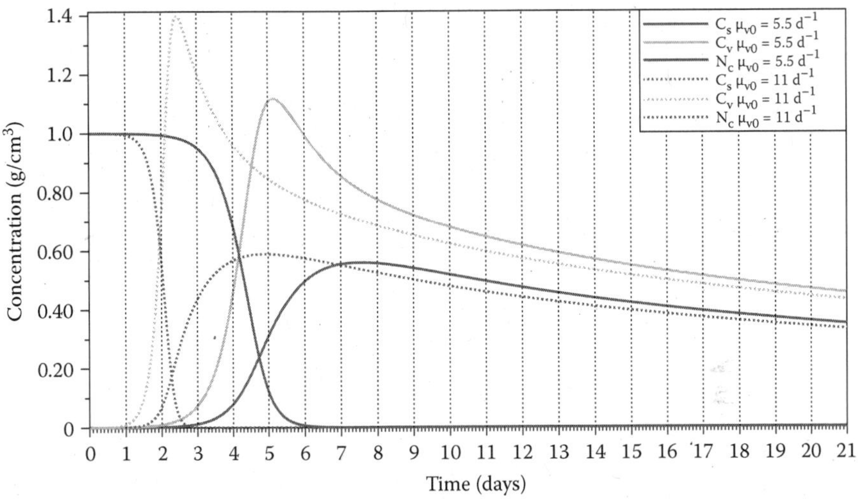

FIGURE 5.9 (See color insert.) Surface concentrations sensitivity to μ_{v0}.

of growth is limited by the available mass of nutrients, the total mass of vegetative growth remains the same.

Figure 5.10 also shows the sensitivity of the model predictions to the rate of vegetative growth, but at the paper–gypsum interface. Comparing Figure 5.10 to Figure 5.9 shows that the predicted behavior is remarkably similar, except that at the interface the graphs are shifted in time.

What is demonstrated by Figure 5.10 is that once the mold is transported to the interior of the board by a mechanism such as diffusion, the rate of growth and

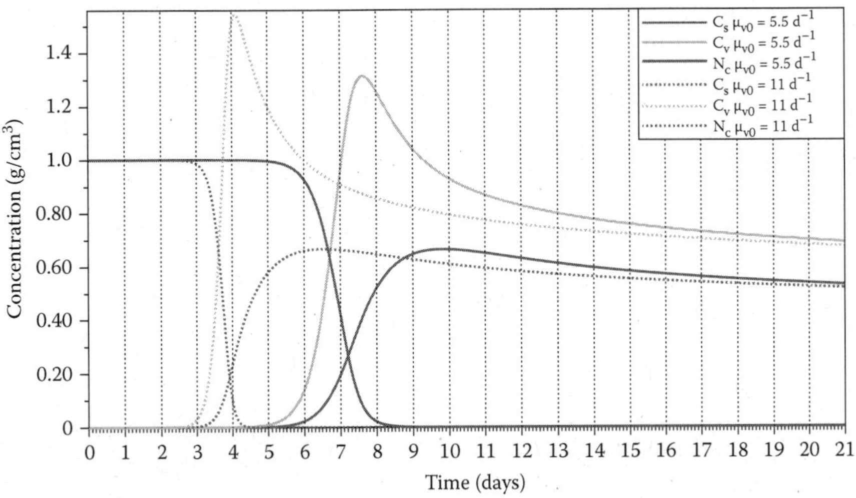

FIGURE 5.10 (See color insert.) Interface concentrations sensitivity to μ_{v0}.

availability of nutrients become the critical parameters. Of course, a more rapid growth nearer the surface does lead to steeper gradients toward the interior, which accelerates inward diffusion, but once a nonzero concentration occurs the vegetative growth rate becomes dominant.

In Figure 5.9 and Figure 5.10 it is clear that the ratio of vegetative and spore concentrations at later times remain the same. Figure 5.11 demonstrates the sensitivity to μ_{s0}. Of course, the absolute value of μ_{s0} is not so much the issue as the ratio of μ_{s0} to μ^1_{v0} , in other words the relative rate of conversion to vegetative growth to the rate of conversion to spores.

As can be readily seen in Figure 5.11, the vegetative concentration plus spore concentration is the same regardless of the value of μ_{s0}, but for the higher value of μ_{s0} a greater fraction of the total mass occurs as vegetative growth than as spores.

Figure 5.12 demonstrates the effectiveness of remediating mold using an agent that kills the vegetative growth. This is represented by using the μ_{vd} term that is zero until remediation starts, and then is set equal to one. The μ_{s0} and μ_{v0} and μ^1_{v0} values remain at 5.5 day^{-1} as in Figure 5.8.

As is dramatically illustrated in Figure 5.12, the vegetative growth is rapidly diminished by the remediation. This is particularly true in the case where remediation is started on day seven in which case vegetative growth concentrations drop almost two orders of magnitude within a day. This can be attributed to the fact that the nutrients were exhausted and therefore the remediation is extremely effective.

In the case where the remediation is started on day four, there are still significant amounts of nutrients and the remediation and growth are competing mechanisms. The remediation agent is sufficiently powerful to overcome growth, although at a rate not as rapid as in the day seven case.

What is of particular interest is the decrease in spore concentrations. The remediation agent does not reduce the spore concentrations per se. Instead, the conversion

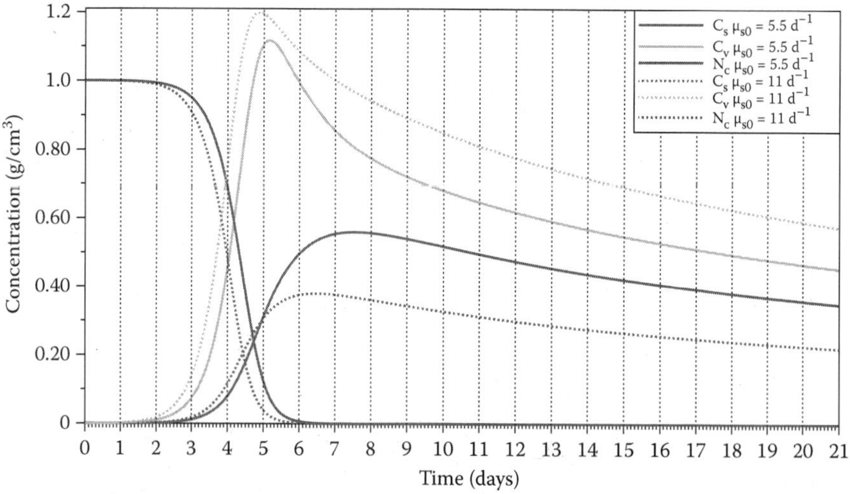

FIGURE 5.11 (See color insert.) Concentrations sensitivity to μ_{s0}.

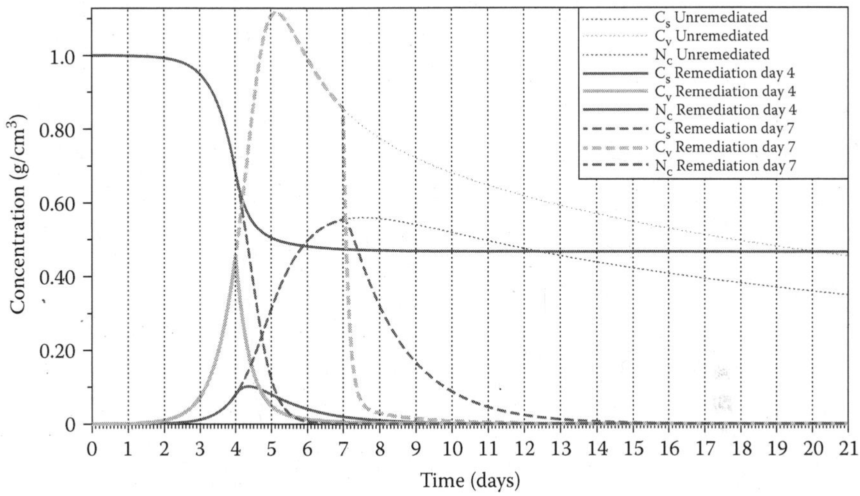

FIGURE 5.12 (See color insert.) Remediating vegetative growth.

from spores to vegetative matter continues at a rate proportional to the spore concentration. The remediation agent is very effective at killing the vegetative matter, so the rate of reduction in the spore concentration is essentially the rate of conversion from spore to vegetative growth.

In particular, it is interesting to note that spores continue to exist in significant concentrations well after the vegetative growth concentrations are below what would be visually observable. It is clear that the remediation needs to be applied in such a way that the remediation agent remains for a sufficiently long period to kill off any new vegetative growth that would form if the agent is removed. This is particularly important in the instance where nutrients remain to fuel vegetative growth.

5.23 CONCLUSIONS: UTILITY OF THE MATHEMATICAL MODEL

The mathematical model describes transport phenomena in regard to mold infestation very well, as explained in the theoretical development and as confirmed by select examples in the numerical analysis. The utility of the model lies in using it to represent existing mold infestation situations in order to provide benefits in a number of ways. For example, one can conduct numerical experiments (in silico, using the model and computer software) to help understand the behavior and interaction between vegetative propagating mold, mold spores, moisture, and temperature variables.

Certainly, if water damage is so extensive, it may be necessary to simply pull damaged drywall and wood, dry the space thoroughly, and then be sure mold is not measured to be significantly present any longer before new wood and drywall are installed. In this case, model analysis is not needed since the situation unfortunately warrants total replacement.

However, for intermediate cases where a leak has occurred, and mold may or may not be evident, the model can be useful, along with analytical data or predictive measures, to determine if the mold can be arrested or perhaps measures can be taken to eliminate at least vegetative-propagating mold, even if spores cannot be safely eliminated (spores are typically very robust and resistant to destructive measures). Certainly, knowing a case history and using a model to depict past events, even if not fully recorded, may be helpful to many workers including consumers, remediation practitioners, and physicians to deal with situations and those affected to determine the best ways to deal with a mold problem.

Therefore, some examples of practical utility for model include the following:

- Use the model to predict how mold infestations may develop in order to prevent them.
- Use the model to describe infestation situations in cases where analytical data do not exist but perhaps where mold is observed and photographs taken.
- Use the model to predict how a mold infestation and products may develop in cases that are largely based on supposition (without much data) as to case scenarios (such scenarios can be developed from the structure of the model and use of historical data from the literature or other experiences).
- Incorporate analytical data into a database and use the model with the database to develop an expert system to predict how mold infestation can develop based on conditions, or "reverse engineer" how a particular mold infestation occurred.
- Use the model to assess the effect of measures to remediate mold such as use of mold growth inhibitors, sporulation inhibitors, or agents to counter formation of chemical and biochemical products of the mold (perhaps catching an infestation in time before structural remediation and "teardown" is required).
- Use the model in research to compare test mold infestations studied in the laboratory or pilot plant to infestations in real-world structures in the field (e.g., see Figures 5.1 to 5.12). This work can be used to improve remediation and help prevent destructive mold situations.
- Apply the model to a time-based scenario on how mold and its products are formed, possibly leading to an existing health problem, by simulating the development of a mold-related health problem that is putatively traced to contamination of a building structure where people are exposed to mold. Such work may help a physician think about what has happened or is happening to a patient by providing better information to understand disease cause. Each type of mold may have its own growth, sporulation, and product formation characteristics with a unique impact on an affected person, physically and mentally.
- Use the model applied to particular mold species and products, and couple findings to the application of therapeutic agents and treatments for a particular mold infection or suspected infection. For example, certain mold infectious spores may penetrate into the lungs based on size and chemical properties and lodge in locales where in vivo vegetative growth can occur, analogous perhaps to how mold grows in dry wall.

- Insurance companies may incorporate use of model findings by retaining practitioners to apply the model to assess the development of mold occurrences in relation to claims.
- Attorneys may incorporate use of model findings by retaining practitioners to use the model to assess legal claims of damages and health problems related to mold infestation.
- Engineers and architects may retain practitioners to use the model to assess new building designs and materials in terms of their adequacy to effectively resist mold infestation.
- Consumers may be able to retain practitioners to use the model to assess an existing infestation situation and also construction defect occurrences.
- All parties can study the model and its description to think about their own particular situations to gain ideas and insight.

REFERENCES

Ashley, P., et al. 2008. Vacuum dust collection protocol for allergens. HUD Office of Healthy Homes and Lead Hazard Control, for HUD Grantees. (Original by Battelle, Version 1, Revised by Quantech, Version 2.)

Bird, R. B., Stewart, W. E., and Lightfoot, E. N. 2007. *Transport Phenomena*, 2nd ed. New York: Wiley.

Buttner, M. P., Cruz-Perez, P., Garrett, P. J., and Stetzenbach, L. D. 1999. Dispersal of fungal spores from three types of air handling system duct material. *Aerobiologia* 15: 1–8.

Buttner, M. P., Cruz-Perez, P., Stetzenbach, L. D., Garrett, P. J., and Luedtke, A. E. 2002. Measurement of airborne fungal spore dispersal from three types of flooring materials. *Aerobiologia* 18(1): 1–11.

Chattoraj, D. K., and Birdi, K. S. 1984. *Adsorption and the Gibbs Surface Excess*. New York: Plenum.

Clarke, J. A., Johnstone, C. M., Kelly, N. J., McLean, R. C., Anderson, J. A., Rowan, N. J., and Smith, J. E. 1998. A technique for the prediction of the conditions leading to mould growth in buildings. *Building and Environment* 34: 515–521.

Covarrubias-Cervantes, M., Mokel, I., Champion, D., Jose, J., and Voilley, A. 2004. Saturated vapor pressure of aroma compounds at various temperatures. *Food Chemistry* 85: 221–229.

Cruz, P., Buttner, M., and Stetzenbach, L. 2001. Specific detection of *Stachybotrys chartarum* in pure culture using quantitative polymerase chain reaction. *Molecular and Cellular Probes* 15: 129–138.

Danchin, A., Medigue, C., Gascuel, O., Soldano, H., and Henaut, A. 1991. From data banks to data bases. *Research Microbiology* 142(7-8) (Sept–Oct.): 913–916.

Delahunty, C., Eyres, G., and Dufour, J.-P. 2006. Gas chromatography-olefactometry. *Journal of Separation Science* 29: 2107–2125.

Doleman, B., Severin, E., and Lewis, N. 1998. Trends in odor intensity for human and electronic noses: Relative roles of odorant vapor pressure vs. molecularly specific odorant binding. *Proceedings of the National Academy of Sciences* 95: 5442–5447.

Gutarowska, B., and Piotrowska, M. 2007. Methods of mycological analysis in buildings. *Building and Environment* 42:1843–1850.

Haugland, R. A., Brinkman, N., and Vesper, S. J. 2002. Evaluation of rapid DNA extraction methods for the quantitative detection of fungi using real-time PCR analysis. *Journal of Microbiological Methods* 50(3): 319–323.

Institute of Medicine. 2004. *Damp indoor spaces and health.* Washington, DC: The National Academies Press.

Michaelis, M., and Menten, M. 1913. Die Kinetik der Invertinwirkung. *Biochemistry Zeitschrift* 49:333–339.

Molz, F. J., Widdowson, M. A., and Benefield, L. D. 1986. Simulation of microbial growth dynamics coupled to nutrient and oxygen transport in porous media. *Water Resources Research* 22(8): 1207–1216.

Sedlbauer, K. 2002a. Prediction of mould fungal formation on the surfaces of and inside building components. Doctoral dissertation, Fraunhofer Institute for Building Physics, Stuttgart, Germany.

Sedlbauer, K. 2002b. Prediction of mould growth by hygrothermal calculation. *Journal of Thermal Envelope and Building Science* 25(4): 321–336.

Wady, L., Bunte, A., Pehrson, C., and Larsson, L. 2003. Use of gas chromatography-mass spectrometry/solid phase microextraction for the identification of MVOCs from moldy building materials. *Journal of Microbiological Methods* 52: 325–332.

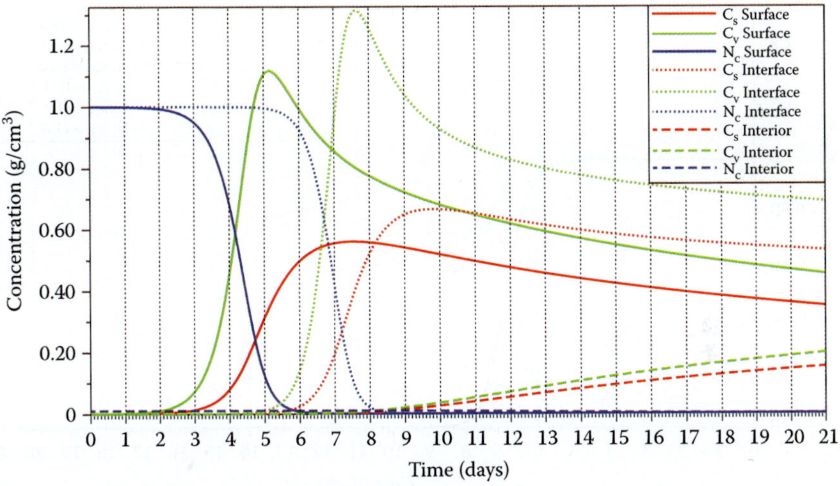

COLOR FIGURE 5.8 Model-predicted concentrations.

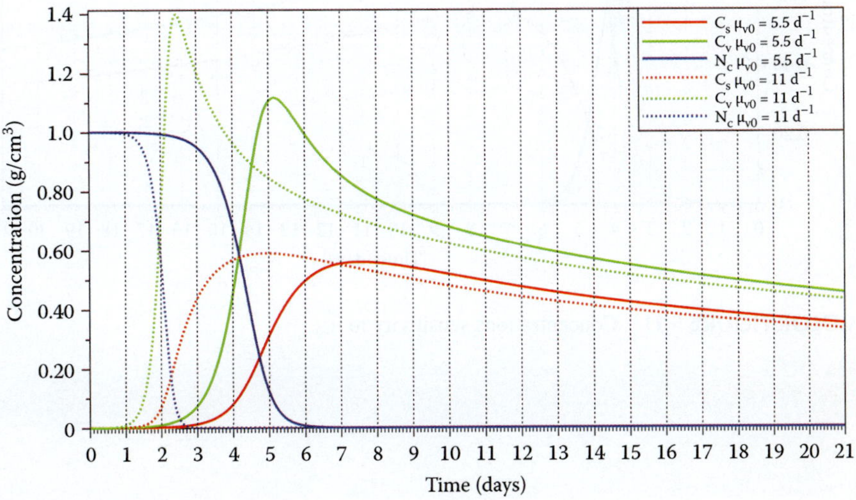

COLOR FIGURE 5.9 Surface concentrations sensitivity to μ_{v0}.

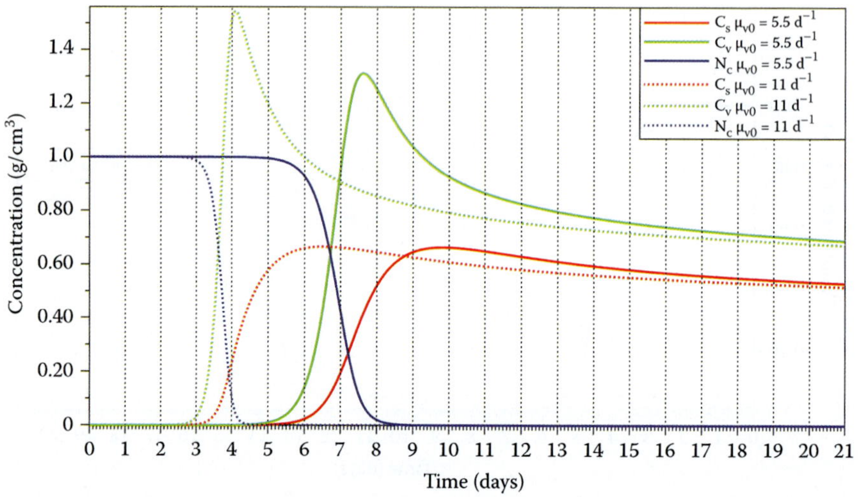

COLOR FIGURE 5.10 Interface concentrations sensitivity to μ_{v0}.

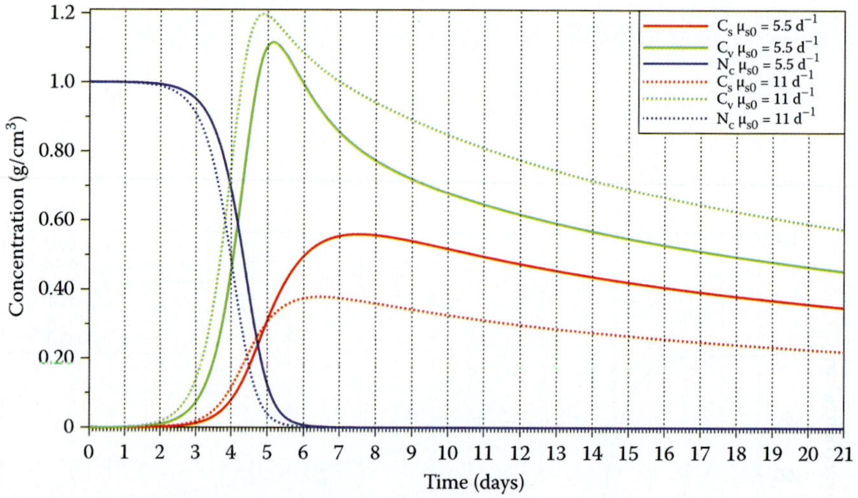

COLOR FIGURE 5.11 Concentrations sensitivity to μ_{s0}.

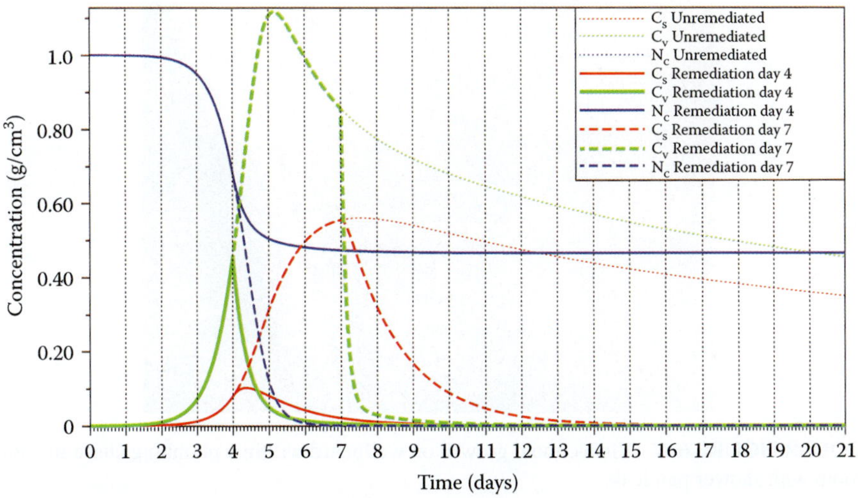

COLOR FIGURE 5.12 Remediating vegetative growth.

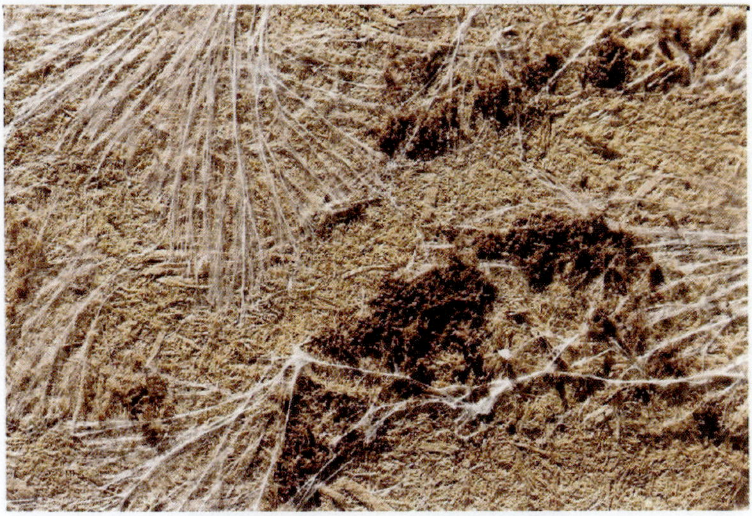

COLOR FIGURE 6.3 Mold growth (rhizomorphs) visible on back side of MWS.

COLOR FIGURE 6.6 Visible mold growth on wallboard within a plumbing chase in bathroom with shower pan leak.

COLOR FIGURE 6.9 Concealed wood rot from a leaking window frame after exterior sheathing was removed.

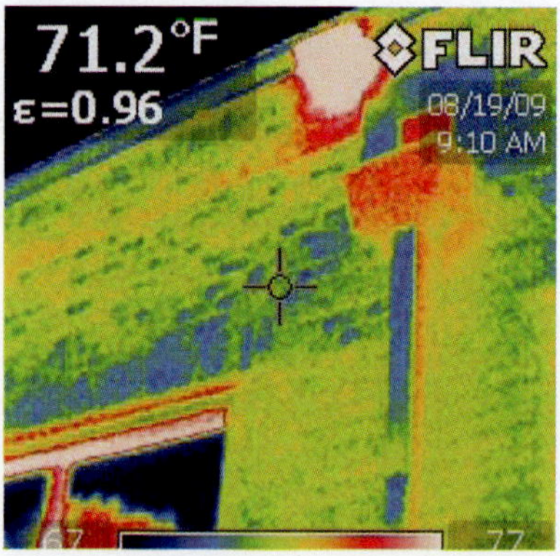

COLOR FIGURE 6.15 Infrared photo of brick wall showing water location.

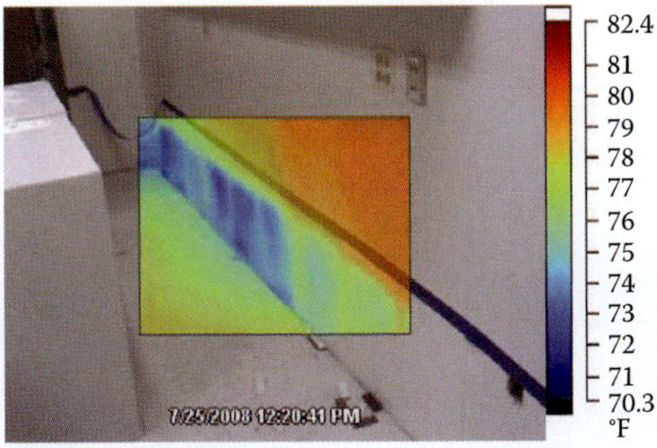

COLOR FIGURE 6.16 Infrared photo showing moisture intrusion into interior wall cavities.

COLOR FIGURE 6.35 The MWS has been removed revealing a failure (biodeterioration) in the underlying building paper.

COLOR FIGURE 6.42 "Water staining," peeling paint, and mold growth at a garden-level bedroom.

COLOR FIGURE 6.44 Removal of interior drywall finish to observe water leakage during testing.

COLOR FIGURE 8.3 *Alternaria alternata* colony on agar.

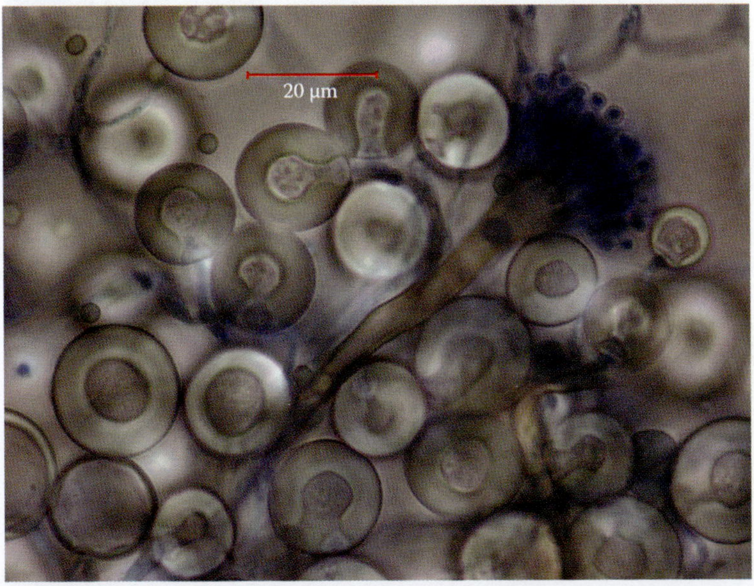

COLOR FIGURE 8.5 Hülle cells and a conidiophre of *Emericella (Aspergillus) nidulans.*

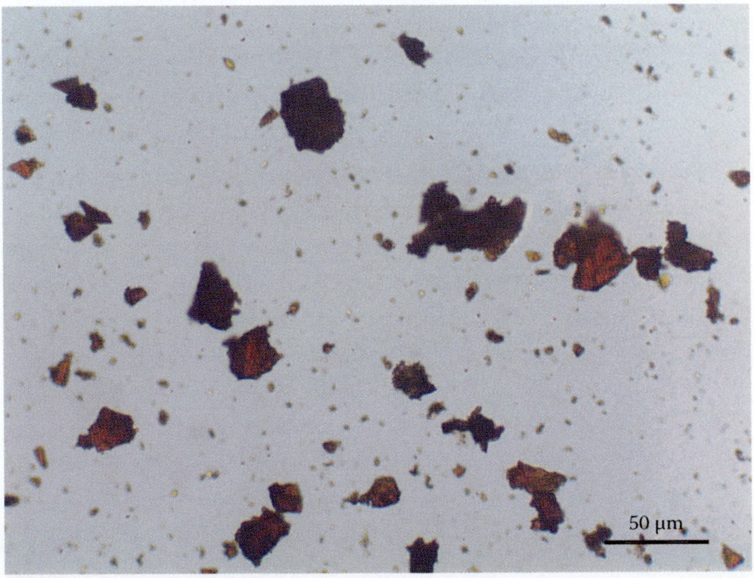

COLOR FIGURE 8.15 Light microscope image of rust flakes.

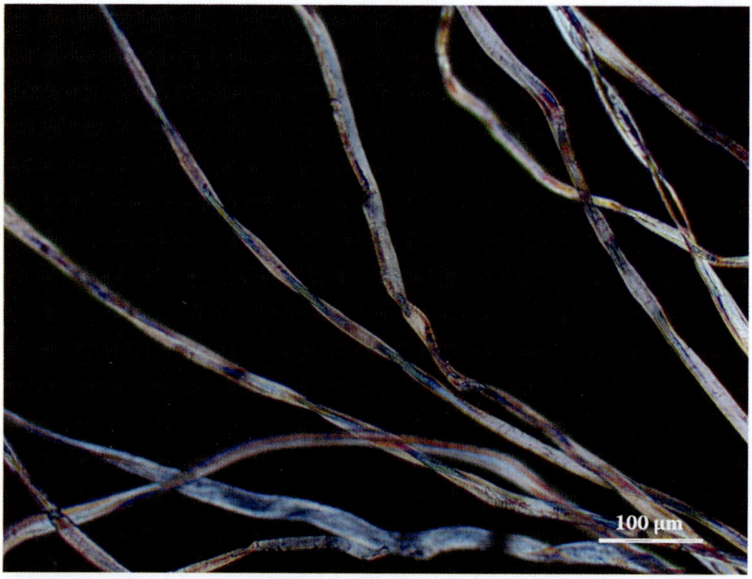

COLOR FIGURE 8.16 Light microscope image of cotton fibers.

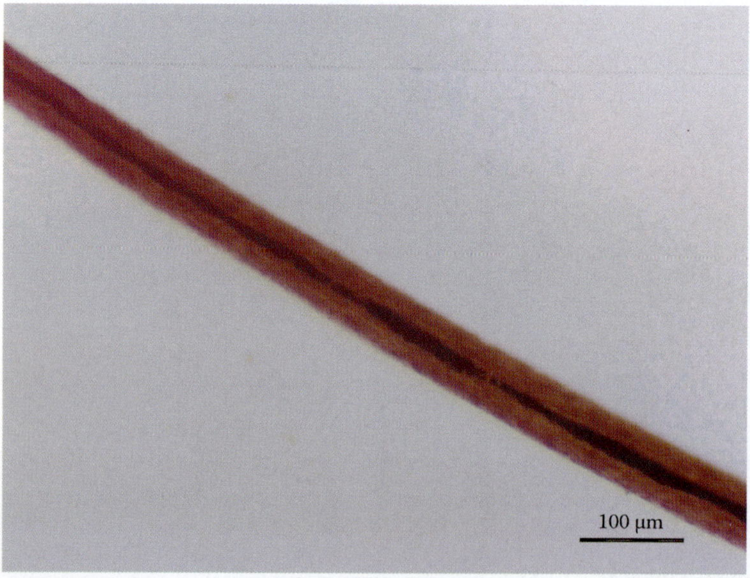

COLOR FIGURE 8.18 Light microscope image of hair.

6 Forensic Studies in Moldy-Damp Buildings

Philip R. Morey, PhD, CIH
Gary N. Crawford, CIH
Michael J. Cornwell, AIA
Brad Caddick, ASP
Tara Toren-Rudisill, AIA
Raoul Webb, PE

CONTENTS

6.1 INTRODUCTION

Buildings or shelters have been erected by man for millennia to provide for both protection and comfort of occupants. It has also been recognized for millennia that the building itself may be a source of indoor contaminants or pollutants. In Books 13 and 14 of the Hebrew Bible, the growth of mold (at that time called "plague") on construction materials such as timber, plaster, and mortar was recognized as "unclean" (Blomquist, 1994; Heller et al., 2003). Moldy construction materials, according to ancient recommendations, were to be disposed outside the city walls and persons from unclean buildings were advised to wash their clothing.

Mold and dampness problems in 21st-century buildings are more complex than those described by Leviticus for a number of reasons including the complex architecture of modern structures. In modern buildings, penetrations through the building envelope whereby water may enter the indoor space, are often numerous and hidden from nondestructive visual inspection. The likelihood of dampness and mold growth conditions in buildings has increased with the widespread use of heating, ventilating, and air-conditioning (HVAC) systems, which are associated with thermal gradients that can cause condensation in some building components.

Forensic investigations in 21st-century buildings are challenging not only because of the architectural complexity of modern structures, but also because of the increasing use of construction and finishing materials susceptible to mold growth. For example, the amorphous cellulose found in products such as paper-faced wallboard, delignified wood siding, protective building papers, and cellulose-containing insulations can become “mold food” when located in damp or wet building niches. Forensic building investigations may be undertaken to look for reasons why leaks occur around window flashing or why water damage occurs from plumbing failures. This chapter provides background and guidance on forensic inspection methods for identification of moisture and mold problems in modern buildings.

During the past 30 years, studies have shown that microbial growth caused by moisture problems can be associated with adverse health effects (Arnow et al., 1978; Hodgson et al., 1987; IOM, 2004; Morey et al., 1984; Samson, 1985). While numerical guidelines (e.g., threshold limit values [TLVs®], permissible exposure limits [PELs]) for mold exposures do not exist, air and surface sampling can be an important part of the forensic inspection process (American Industrial Hygiene Association [AIHA], 2005, 2008). Several examples where microbial assessments were an aid in forensic studies are provided in this chapter.

6.2 BIODETERIORATION OF MANUFACTURED/ENGINEERED/HARDBOARD WOOD SIDING

A building was constructed with manufactured wood siding (MWS), which had been painted white. The construction materials located beneath the MWS consisted of an exterior sheathing of paper-faced gypsum wallboard (GWB), an asphaltic building paper on the outboard surface of the GWB, and wood framing beneath the GWB. Within a year or two after construction, it was observed that the wood siding was becoming discolored by mold growth. Macroscopic mushroom-like structures were observed growing out from the painted siding in some areas (Figure 6.1 and Figure 6.2). The back, inboard surface of the wood siding was extensively covered with a mycelial growth (rhizomorphs; Figure 6.3). It was visually apparent that the wood siding was undergoing biodeterioration caused by entry of water into the building envelope. Common causes of water damage to exterior MWS siding include the following:

FIGURE 6.1 Discoloration and mold growth visible (arrows) on outer surface of MWS.

- Improper manufacture—The manufactured products are typically created using wood strands or fibers that are coated with a resin binder and compressed to simulate the density of natural wood. The boards are then coated with a moisture-resistant overlay. Some more recent products are also treated with zinc borate to aid in resistance to termites and rot. Some early products were found to be insufficiently compressed and contained overlays with insufficient water-penetration resistance or insufficient coverage.

FIGURE 6.2 Close-up view of discoloration and mold growth (arrow) on surface of MWS.

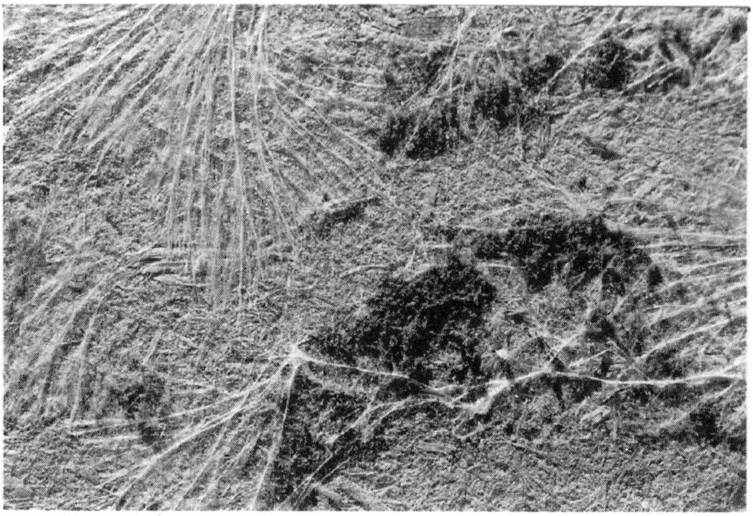

FIGURE 6.3 (See color insert.) Mold growth (rhizomorphs) visible on back side of MWS.

- Improper installation—Large gaps between components can allow excess water to enter and accumulate between the back of the siding and the weather barrier over the sheathing. Any deficiencies in coating or density can then cause swelling, bucking, and fungal growth. The failure to seal around penetrations, such as windows, doors, exterior lighting, and hose bibs, can also contribute to excessive water entry and resultant damage.
- Improper maintenance—The failure to maintain protective coatings, sealants, and repair any impact damage will also accelerate moisture exposure and resultant damage.

Typical damage can consist of swelling, buckling, warping, cracking, fungal growth, and eventual deterioration. Air sampling for culturable molds was carried out in one occupant apartment where the envelope construction materials were heavily impacted by biodeterioration and as a control in the outdoor air. The analytical results of sampling are presented in Table 6.1. The molds present in the outdoor air were dominated by *Cladosporium cladosporioides* (common outdoor-sourced mold that grows on leaf surfaces) with minor amounts of *Aspergillus* and *Penicillium* species. By contrast, the room air in the affected apartment was dominated by *Penicillium* and *Aspergillus* species (e.g., *A. niger* and *P. brevicompactrum*) as well as yeasts, which are fungi that can grow on wet biodegrading construction materials. The data in Table 6.1 suggest that fungi present on moldy envelope construction materials (e.g., siding) were entering the room air, at least in the apartment where sampling was carried out.

Moldy construction materials within the building's envelope were removed and replaced with new construction materials with attention given to prevention of water entry into the building envelope. The building remained occupied during the remediation process. The air within occupied apartments was positively pressurized during

TABLE 6.1
Comparison of Culturable Molds in Indoor Air in Apartment with Moldy MWS and in the Outdoor Air

Taxa	CFU/M³
Rank Order Concentration of Molds in Indoor Air	
Asp. niger	2,080
Pen. crustosum	800
Pen. brevicompactum	160
Yeasts	6
Clad. cladosporioides	4
Rank Order Concentration of Molds in the Outdoor Air	
Clad. cladosporioides	150
Pen. brevicompactum	6
Alternaria alternata	6
Pen. chrysogenum	4

Note: Malt extract agar; N = 6 for indoor and outdoor air.

the remediation process so as to reduce the inward movement of mold spores into indoor air of the apartments.

The following actions can be effective in preventing water damage and biodeterioration of MWS used in envelope construction:

- The first consideration is to determine the manufacturer of the MWS and if the product has been recalled due to manufacturing defects. If the product was subject to recall, the product should be replaced.
- If the product is determined to be in sound condition, then repainting, replacement of failed or missing sealants, and repair or replacement of any damaged wall components should suffice.

6.3 BIODETERIORATION IN WALL ASSEMBLIES ASSOCIATED WITH PLUMBING LEAKS: A HOTEL EXAMPLE

While plumbing leaks can occur in any building, the frequency of leaks and the severity of water damage and biodeterioration is often greatest in hotels. Plumbing failures in hotels occur for a variety of reasons including poor design; inadequate maintenance under conditions of near 100 percent occupancy; and the vast number of potential failure points associated with piping, valves, and water outlets. Hidden leaks are the most problematic because the water damage and consequential mold growth can become extensive over long time periods before the damage becomes visually evident in occupied spaces.

In the following example, leakage was traced to a defective waterproofing membrane in shower pan assemblies in a new hotel. Guests complained about moldy odors in room air, wet carpet, and moldy paperfaced wallboard on walls adjacent

FIGURE 6.4 Chronically wet carpet adjacent to a bathroom wall suggests the presence of a plumbing leak.

to bathrooms (Figures 6.4 to 6.6). Walls containing plumbing and utility lines were carefully opened (high-efficiency particulate air [HEPA] vacuum used to control dust emissions; see destructive testing procedures in Section 6.4) and visible mold growth was found on the wallboard lining the wall cavities/pipe chases. By a process of elimination, the primary source of water leakage in the hotel was determined to be defective waterproofing membranes beneath shower assemblies. A waterproofing membrane is installed in the lower part of the shower stall covering the floor and

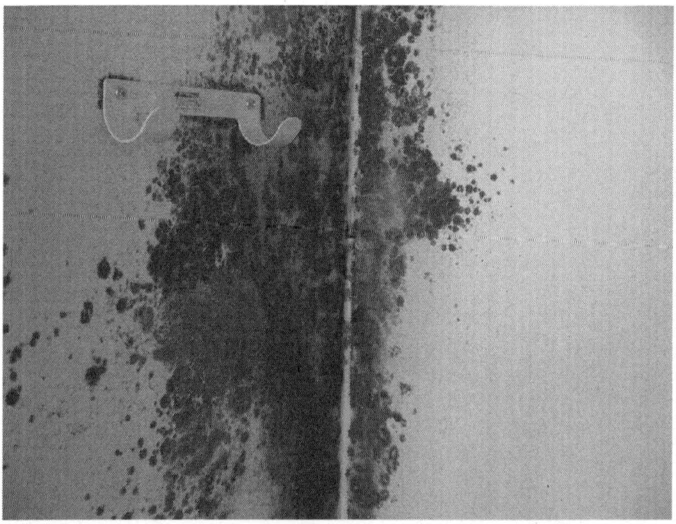

FIGURE 6.5 Mold growth occurred on wall adjacent to shower stall.

FIGURE 6.6 (See color insert.) Visible mold growth on wallboard within a plumbing chase in bathroom with shower pan leak.

running upward about 6 inches in the walls in order to prevent the capillary movement of water into masonry and wall infrastructure.

A flood test was carried out in a shower pan with a defective waterproofing membrane (Figure 6.7 and Figure 6.8). Water from the shower had entered adjacent wall systems so that the moisture meter inserted into the wall gave a maximum water saturation reading.

FIGURE 6.7 Shower with standing water (arrows) in pan during a shower pan waterproofing membrane test.

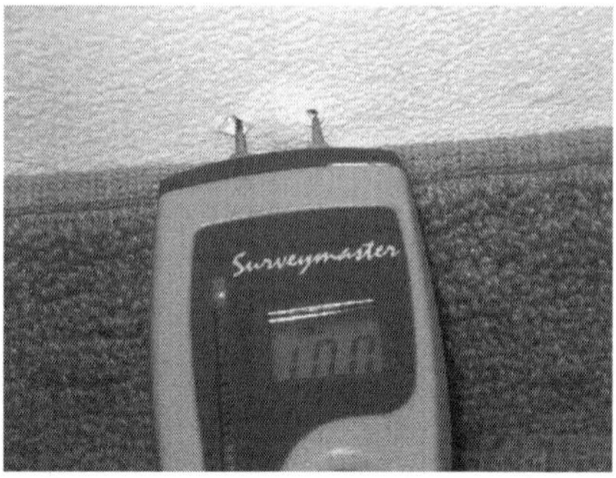

FIGURE 6.8 Moisture measurement of wall behind shower during shower pan water test. The waterproofing membrane was defective.

The hotel inquired about the exposure risk to people who might be present in rooms with water and mold damage. Air sampling (samples) carried out in one guest room showed that 99 percent of the collected fungal structures were *Penicillium-Aspergillus* spores with an average concentration of 48,000/m^3. The outdoor air (samples collected on the roof) was dominated by basidiospores and *Cladosporium* with *Penicillium-Aspergillus* spores comprising only about 6 percent (or 100s/m^3) of the total count. While adverse health effects cannot be predicted by environmental air sampling, it was clear that the air in guest rooms was degraded (see Miller et al., 2000; Morey et al., 2003) and that room occupants, especially housekeepers, would be at the highest risk.

Specific actions useful in avoiding waterproofing membrane failures in showers include:

- Verify that there are no separations and discontinuities in the membrane.
- The drainage line passing through the membrane must be sealed.
- Prior to occupancy, verify that the membrane is leak proof when the shower pan is filled with water.
- The substrate under the pan should be sloped so that the pan itself slopes toward the drain.
- The weep holes for the drain neck should be clear, and the weep hole area around the drain neck should be surrounded with P-rock to properly ensure long-term drainage out of the pan area.
- In the case of flexible pan membranes, the corners should be properly folded so that all crease troughs terminate at the pan rim level.

General actions that should be considered to minimize plumbing leaks include:

- Allow adequate time to test plumbing lines for leaks when wall systems are open.
- Avoid placing water valves in locations above ceiling spaces so as to minimize damage to ceiling materials.
- Use moisture-resistant materials (e.g., concrete board, fiberglass wallboard) to line chases with piping that may be subject to water leaks.

FIGURE 6.9 (See color insert.) Concealed wood rot from a leaking window frame after exterior sheathing was removed.

6.4 DESTRUCTIVE INSPECTION TO FIND HIDDEN STRUCTURAL DEFECTS

The building envelope must be designed, constructed, and maintained in a manner that effectively protects the structure from moisture-related damage. Walls, wall penetrations, and roofs must shed water and manage moisture that unavoidably enters the building envelope. Failure to effectively shed and manage moisture usually leads to degradation of building materials and indoor air quality. A thorough inspection means finding the defects that are often not apparent on finished, exposed surfaces. Interior and exterior surfaces of walls and roofs often appear normal and unaffected yet concealed defects and damage could be extensive (Figure 6.9 and Figure 6.10).

FIGURE 6.10 Vinyl wall covering was concealing abundant mold growth.

FIGURE 6.11 Pin-type moisture meter reading of wood water content.

The concept of destructive inspection of building elements is often met with resistance from building owners and managers. The reasons are typically:

- Concern about resulting damage
- Cost and effort required to repair materials impacted by destructive inspection
- Disruption of normal use and occupancy
- The unsightly appearance created
- Fear that odors and contaminants will be released and potentially affect occupant health

The resistance to conducting destructive inspections can usually be overcome by first preparing and presenting a well-conceived investigative protocol. The initial steps are typically a thorough, nondestructive inspection of accessible building elements and surfaces supplemented by nondestructive test methods such as the use of moisture meters and infrared cameras to detect concealed moisture. These techniques can often identify concealed moisture and either eliminate the need for destructive inspection or focus on limited areas to minimize the number of representative openings needed and to pinpoint water leakage locations (Figures 6.11 to 6.16).

FIGURE 6.12 Radio frequency (RF) moisture meter reading of masonry moisture content.

FIGURE 6.13 RF moisture meter reading concealed water content in EIFS.

FIGURE 6.14 Photo of brick wall with suspected moisture infiltration.

FIGURE 6.15 (See color insert.) Infrared photo of brick wall showing water location.

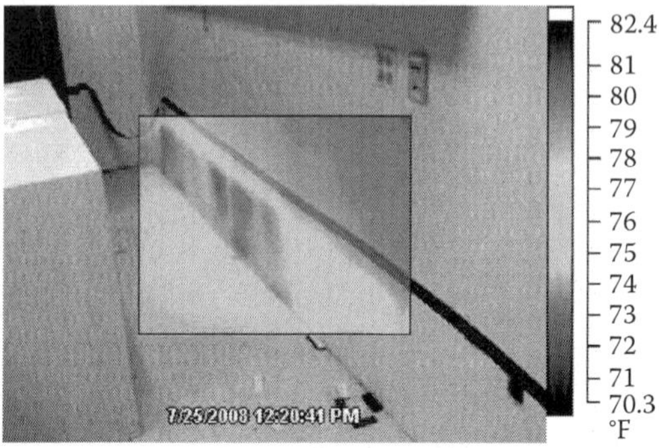

FIGURE 6.16 (See color insert.) Infrared photo showing moisture intrusion into interior wall cavities.

Another destructive investigation technique is to look for visible clues that can be indicative of potential concealed damage. As an example, the appearance of efflorescence on a masonry wall usually is a telltale sign of water infiltration into wall materials (Figure 6.17 and Figure 6.18).

Another relatively nondestructive technique used to locate moisture intrusion and potential mold problems is to pull back carpeting along walls and examine the carpet

FIGURE 6.17 Efflorescence in this section of a brick wall was an indication that moisture may be entering the wall cavity in this area.

FIGURE 6.18 The destructive inspection of the corresponding interior wall (see Figure 6.17) revealed extensive mold and rot in the plywood sheathing. Water infiltration was then traced to improperly installed coping and flashing in the parapet.

FIGURE 6.19 Removal of carpet reveals previous moisture intrusion.

tack strip, padding, subfloor, and backside of the carpet for stains or mold, which should help reduce the number of locations where destructive wall openings need to be made. Carefully pulling back base trim molding can also provide valuable clues as to where moisture is entering the wall system thus reducing the number and size of destructive openings needed to diagnose wall cavity conditions and water infiltration sources (Figures 6.19 and Figure 6.20).

After the nondestructive inspection and tests are completed, the investigator should develop one or more hypotheses about where and how moisture may be entering building elements and where concealed mold and degraded building materials are most likely located. Evidence from this initial investigation should help convince the building owner or manager of the need for a destructive inspection to prove or disprove the hypotheses. Decisions need to be made on whether openings in the building envelope should be made from the inside or outside. Factors in that decision involve technical needs for information balanced against disruption, geometry, and location of damage; contamination concerns; and ease of repair factors.

An advance agreement should be in place between the party conducting the destructive inspection and the owner or manager regarding who will make the openings and repair or restore the affected building materials. Often, the damage is determined to be extensive by multiple, strategically located inspection openings. The

FIGURE 6.20 Pulling back base trim molding reveals previous water infiltration.

FIGURE 6.21 These figures (see also Figure 6.22) show destructive inspection openings in a brick wall that reveal improperly installed flashings and weeps, which were allowing water infiltration into wall cavities.

identification of deteriorated materials at many such locations in an affected large area usually leads to a systematic dissection of large areas to observe the water markings for color, geometry, and affected material symptoms such as texture. Careful observation of these symptoms will define the cause and severity of the water intrusion. If it is an exterior masonry wall that is to be opened, it may require masonry and carpentry contractors to make the inspection and dissection openings, and then repair the wall when the work is complete. A similar consideration should be given to using a roofing contractor to make and repair roof inspection openings immediately after the inspection to preclude additional damage (Figures 6.21 to 6.24).

FIGURE 6.22 These figures show destructive inspection openings in a brick wall that reveal improperly installed flashings and weeps, which were allowing water infiltration into wall cavities.

FIGURE 6.23 A small inspection opening was made below a window having suspected window frame leakage. Due to the shape and extent of the apparent underlayment damage, it was decided to enlarge the inspection opening (see Figure 6.24).

FIGURE 6.24 Expansion of the opening in Figure 6.23 was appropriate. Most of the wall has been compromised by voluminous or long-term water intrusion, material degradation, and mold. This entire wall section was ultimately demolished and remediated. This apartment was occupied and remained so throughout remediation, against urgent advice.

FIGURE 6.25 Moisture meter shows area of water saturated substrate.

The use of a nondestructive moisture meter can help locate areas of potential moisture infiltration and construction defects. This technique guides the destructive inspection to the most information productive locations (Figure 6.25 and Figure 6.26).

When destructive openings need to be made within the building there is an increased concern about potential contamination issues and exposure of people to bioaerosols. Preplanning destructive inspections from the building interior should, at a minimum, include the following considerations:

- Is the building occupied or vacant?
- Are there occupants with health risks?
- Are there furnishings, merchandise, equipment, and so forth that could be adversely impacted?
- Can the occupants vacate the area or building during the destructive inspection?

FIGURE 6.26 Opening cut into exterior wall showed water damage and mold growth traced to a leaking window frame installation.

Since many interior destructive inspections involve less than 10 square feet of building material being disturbed, elements of the New York City Mold Remediation Guidelines (New York City Department of Health and Mental Hygiene, 2008) may be applicable to inspection work as well. The following are excerpts from that publication:

1. Efforts should also be made to minimize the generation and migration of any dust and mold.
2. The work area should be unoccupied.
3. If work may impact difficult-to-clean surfaces or items (e.g., carpeting, electronic equipment), the floor of the work area, egress pathways, and other identified materials/belongings should be removed or covered with plastic sheeting and sealed with tape.
4. Efforts should be made to reduce dust generation. Dust suppression methods, particularly during any cutting or resurfacing of materials, are highly recommended. Methods to consider include:
   a. Cleaning or gently misting surfaces with a dilute soap or detergent solution prior to removal;
   b. The use of high-efficiency particulate air (HEPA) vacuum-shrouded tools; or
   c. Using a vacuum equipped with a HEPA filter at the point of dust generation.
   d. Work practices that create excessive dust should be avoided.

If the interior destructive inspection will affect more than 10 square feet of building materials then additional elements of the New York City Mold Remediation Guidelines (New York City Department of Health and Mental Hygiene, 2008) may be advisable. The following are excerpts from that publication:

1. The HVAC system servicing this area should be shut down during work.
2. Isolate the work area using plastic sheeting sealed with duct tape.
3. Furnishings should be removed from the area.
4. Ventilation ducts/grills, any other openings, and remaining fixtures/furnishings should be covered with plastic sheeting sealed with duct tape.
5. Consider using an exhaust fan equipped with a HEPA filter to generate negative pressurization.
6. Egress pathways should also be covered.
7. The work area should be unoccupied.

6.5 EXAMPLE OF DESTRUCTIVE INVESTIGATION OPENINGS TO DISCOVER CONCEALED STRUCTURAL ELEMENTS

Destructive investigation openings are often a critical part of a building inspection. These openings allow forensic architects and engineers to observe concealed conditions and to better understand the details of a building's construction.

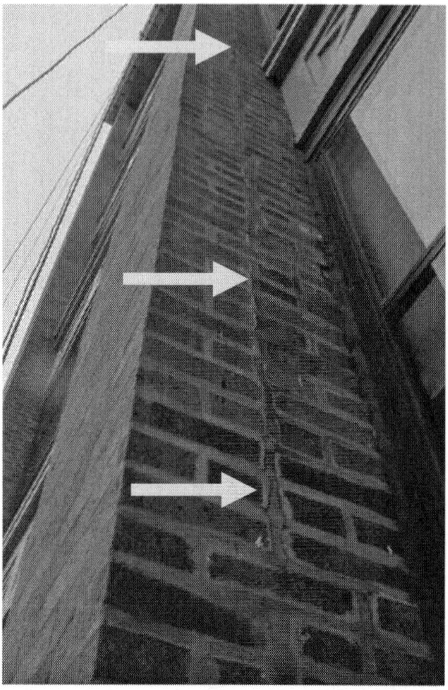

FIGURE 6.27 Extensive vertical cracking (arrows) in masonry cladding.

The study building is a historic structure with a variety of structural components. Masonry load-bearing walls exist at three of the exterior elevations and a steel frame structure supports the fourth exterior wall. A heavy timber structure supports the building's interior spaces. During a façade inspection, a forensic architect or engineer observed distressed masonry cladding adjacent to the steel structure (Figure 6.27). Destructive investigation openings were created for the forensic architect or engineer to inspect the condition of the concealed steel structure of the wall. Once the inspection openings were observed and the extent of the deterioration analyzed, an appropriate repair program was developed.

Poor maintenance of exterior cladding components leads to their deterioration (i.e., failed mortar joints in a masonry wall), which allows an excessive amount of moisture to be absorbed into the wall. Unprotected steel exposed to moisture corrodes. As steel corrodes, it can expand up to 10 times its original dimension. The buildup of corrosion by-products produces expansive forces in the masonry cladding, which causes distress in the masonry, typically appearing as cracks, spalls, or displacements in the masonry. These conditions of distress create additional locations for water to penetrate the exterior walls, which results in an accelerated deterioration of mortar joints and further corrosion of the steel (Figures 6.28 to 6.34).

FIGURE 6.28 Destructive inspection opening to observe concealed conditions.

FIGURE 6.29 Destructive inspection opening to observe concealed conditions.

FIGURE 6.30 Larger inspection opening required to determine extent of deterioration.

FIGURE 6.31 Larger inspection opening required to determine extent of deterioration.

FIGURE 6.32 Complete failure of structural components (arrow).

FIGURE 6.33 Severe corrosion (arrow) of structural components.

FIGURE 6.34 Severe corrosion of structural components. Metal supports have deteriorated (arrows).

6.6 BIODETERIORATION OF BUILDING PAPER

A component of building envelope construction that has historically been considered important has been asphaltic paper ("building paper"), typically located in the exterior portion of the wall. Building papers are designed to be hydrophobic and shed liquid water that may enter the envelope from outside sources such as rain ("bulk" water). Water that enters the building envelope and impacts the outboard surface of the protective building paper is supposed to flow downward (by gravity) and exit the envelope through drain holes (weeps) at the base of the wall.

However, it has been shown in all cases that standard building paper, such as "Grade D" or "# 15 felt," which has been exposed to continuous high-volume bulk water, and/or extended time bulk water or entrapped high-moisture presence has failed. Paper water transmission tests and mil spec tests referenced in the codes require only a single water exposure, with the tests reporting the time to saturation of a single test cycle, rather than repeated tests to determine cyclic or long exposure time performance. In view of the figures in this chapter and the volume of related lawsuits associated with dramatic structural failures, and mold and water intrusion caused by such performance, the extended performance tests mentioned can be considered to have been done empirically.

Caution in the use of building wraps is similarly warranted. Most wraps are woven polymers that are often adversely affected (and therefore limited) in their installation, which involve stretching or forming around corners such as is done where they are used as "window sill flashing." Water can readily penetrate these wraps where their weaves are deformed and their surfaces come into intimate contact with papers, thereby defeating their waterproofing capability (the "tent" effect). The reason for these behaviors is due to the reduced surface tension of the wrap, which is required for waterproofing performance.

FIGURE 6.35 (See color insert.) The MWS has been removed revealing a failure (biodeterioration) in the underlying building paper.

The outer and inner surfaces of building papers as revealed by destructive inspections may become biodeteriorated. Biodeterioration of building paper may occur in building envelopes with various claddings such as cementateous stucco or MWS. Also, noteworthy is the appearance of wrinkling of building papers in areas affected by entry of bulk water (Figure 6.35 and Figure 6.36). This wrinkling was caused by

FIGURE 6.36 The inner portion of the envelope wall has been removed revealing mold growth (arrows) on the inner surface of the building paper.

diffusion of moisture away from the bulk water location to less moist areas. Some visible mold growth (mostly *Stachybotrys* and *Chaetomium*) can occur on the inside face of the paper. Although building papers are impregnated with hydrophobic materials (e.g., wax, asphalt), these materials are very susceptible to water volume and exposure time; and the cellulose within the paper is susceptible to deterioration caused by mold growth.

Architects are now beginning to follow the protocols adopted by remediators: design buildings with composite asphaltic/polymeric "peel-and-stick" material applied to wallboard or MWS, followed by a layer of #15 felt or building wrap (and self-furring lath for stucco walls). To form a "drain plane" in the stucco example, the felt/wrap layer would be furred away from the peel-and-stick; however, the drain plane is still only rarely done in most hot and humid areas.

There are various specifications of building papers, according to Federal Specification UU-B-790a:

- 4 Types
- 4 Grades
- 12 Styles

The paper types are

- Type I—Barrier Paper
- Type II—Concrete Curing Paper
- Type III—Fire Resistant
- Type IV—Insulation Tape Paper

The primary focus in this work is the water characteristics of barrier papers. There are many other building papers listed in the Federal Specification, but discussion of these is beyond the scope of this work. Type I papers include

- Grade A—High water-vapor resistance
- Grade B—Moderate water-vapor resistance
- Grade C—Water resistant
- Grade D—Water-vapor permeable

Grade D paper has been the most widely used in the past, followed by #15 felt. Asphalt-saturated Kraft paper (Grade D) is available in one- or two-ply rolls. The two-ply rolls are rated at 10, 30, or 60 minutes. However, the minimum requirements for Grade D paper are

- Dry tensile strength: 20 lbs both machine and cross-direction
- Water resistance: 10 minutes
- Water vapor transmission
  - Maximum: No limit
- Minimum: 35 grams

Note that roofing felt (commonly known as "15 lb" or higher) consists of multiple layers of loosely laid cellulose fibers. The fiber source is unbleached softwood pulp and a high percentage of recycled fibers. This results in felt having less uniform fiber structure, larger pore size, and bulky matrix.

Building codes depend upon American Society for Testing and Materials (ASTM) guides as the technical bases to support their respective topics. In the case of building papers, product standards are addressed by

- ASTM D-779: Water Resistance of Paper, Paperboard, and other Sheet Materials by the Dry Indicator Method (Boat Test)
- ASTM D-828-93: Tensile Properties of Paper, Paperboard using Constant Rate of Elongation Apparatus
- ASTM E-96-95: Water Vapor Transmission of Materials

These ASTM guides, together with Federal Specification UU-B-790a, form a set of living design and utilization documents for building papers. It is interesting to note that the Federal Specification has survived in a relatively unchanged state since its inception in the early 1960s. Unfortunately, these documents address only a single water exposure event in their testing scenarios.

As proven by every building forensic project involving building papers, regardless of the kind of building paper specified for an envelope wall, the paper is not expected to resist chronically wet conditions, nor repeated wetting and drying. Architects and engineers must understand and accept that a number of specific moisture problems are often associated with building paper failure (biodeterioration).

6.7 EXTERIOR INSULATION AND FINISH SYSTEMS (EIFS) AND DRAINAGE PLANES

The outer part of the envelope in some buildings may be composed of an exterior insulation and finish system (EIFS), which is a thick layer of foamboard overlayed by a thin cementitous layer. Immediately beneath the foamboard, an exterior sheathing is found, which may consist of paper-faced GWB or particle board. Water can enter the EIFS structure, penetrate beneath the foamboard, and cause the mold growth because it (the water) then cannot drain out of the envelope construction (there was no air space between the foamboard and the GWB, which would allow water to drain out of the envelope).

Mold growth can occur on biodegradable sheathing in EIFS construction (Figure 6.37). *Stachybotrys* is commonly found on water-damaged cellulose-containing sheathing and GWB (Figure 6.38). *Penicillium* and *Aspergillus* species also grow on water-damaged sheathing and GWB.

Table 6.2 shows air sampling data for culturable molds found in one apartment in an EIFS clad building. *Penicillium chrysogenum* dominated the mold spores found in the indoor air in the apartment, while *Cladosporium cladosporioides* dominated the molds in the outdoor air. This kind of air sampling data suggests

FIGURE 6.37 Destructive inspection of EIFS showing foamboard(s) and mold growth (arrows) on underlying exterior sheathing.

that biodeterioration and mold growth may be occurring in the envelope wall of the sampled apartment.

Historically, most EIFS have been "barrier" systems, that is, a single line of defense against water intrusion. The barrier consists of the reinforced cementitious base layer, perimeter sealants, and any related flashings. Any breach in the components will allow water to enter behind the barrier and come into contact with other wall components.

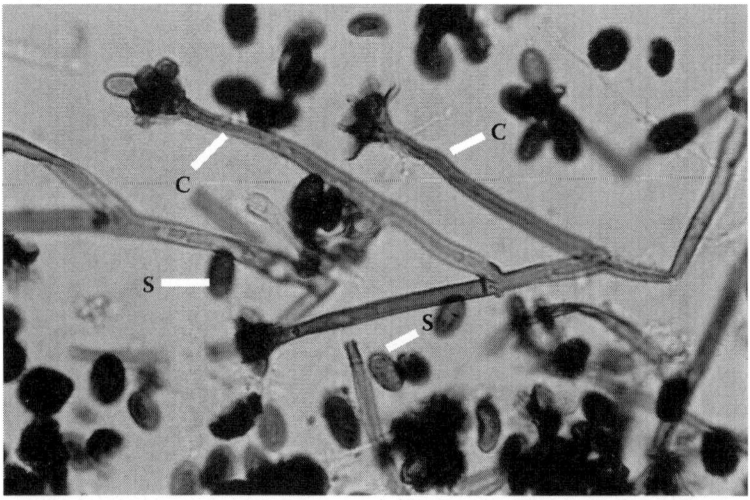

FIGURE 6.38 *Stachybotrys* spores (s) and *condiophores* (c) present on moldy exterior sheathing.

TABLE 6.2
Comparison of Three Kinds of Molds in the Indoor Air in a Building with Moldy EIFS and in the Outdoor Air

Taxa	CFU/M³
Rank Order Concentrations (CFU/M³) of Molds in Indoor Air	
Penicillium chrysogenum	210
Penicillium brevicompactum	73
Cladosprodium cladosporioides	58
Rank Order Concentration of Molds in the Outdoor Air (CFU/M³)	
Cladosprodium cladosporioides	321
Penicillium brevicompactum	16
Penicillium chrysogenum	13

Note: Malt extract agar; N = 3 for indoor and outdoor air.

The most common installation defects are as follows:

- Perimeter sealant applied to the textured finish layer instead of the cementitious base coat.
- Failure to install the required horizontal control joint at the second floor line of wood frame structures. This joint is needed to allow for cross-grain shrinkage of wood components concentrated at the second floor line without buckling the EIFS.
- Failure to install roof "kick-out" flashings at the base of the flashing at the transition between the one-story portions of a structure and the two-story walls above.
- Failure to install sealants at penetrations of the barrier.
- Failure to install the required reinforcing mesh at window corners and other manufacturers' recommended locations.
- Failure to install window and door flashings.

Due to past installation defects, new systems have an added "drainage plane" and weep-screened accessories so that any water penetrating the face barrier is contained and allowed to weep (drain) from the system without damaging the wall components or leading to fungal growth. The "drainage systems" are especially important when EIFS is utilized over wood frame construction or sheathings, which are sensitive to moisture exposure.

6.8 LEAKS ADJACENT TO PENETRATIONS THROUGH THE BUILDING ENVELOPE

Water leakage adjacent to window, door, and other penetrations through the building envelope can occur for a variety reasons, including inadequate design, improper installation methods, and insufficient maintenance of the building's exterior enclosure.

Water leakage manifests itself on interior finishes in a variety of manners, but typically includes bubbling or peeling paint, warped wood components, or "water staining."

Forensic architects and engineers often use controlled water testing onsite to simulate a rainfall to determine the nature of the construction defect(s) that may result in moisture entry around openings in the building envelope. Simulated water intrusion testing can be performed using a variety of standardized tests. Some of these tests include:

- ASTM E 1105—Standard Test Method for Field Determination of Water Penetration of Installed Exterior Windows, Skylights, Doors, and Curtain Walls, by Uniform or Cyclic Static Air Pressure Difference
- AAMA 501.2-03—Quality Assurance and Diagnostic Water Leakage Field Check of Installed Storefronts, Curtain Walls, and Sloped Glazing Systems

Although each standardized test is somewhat different, the general concept remains constant. The area of testing is determined based on field observations, the test area is prepared in a controlled manner, water is applied at a specific pressure for a predetermined amount of time, and the various construction elements within the test area are methodically tested to identify and then pinpoint the exact source(s) of the leak(s). Removal of interior finishes (GWB, plaster, etc.) is typically required to observe water leaks during testing.

Once the cause(s) of water leaks are determined, the forensic architect/engineer can then design repairs. Upon completion of repairs, water tests should again be performed in order to verify the effectiveness of the repairs and to determine if any other leaks exist. Many times, leaks will be caused by multiple sources and only after the primary source is repaired can the secondary source be identified.

Some common kinds of defects that occur in envelope construction and recommendations to fix the defects are as follows (Figures 6.39 to 6.50):

- Failed sealants at penetration locations
- Improper flashing around window and door openings
- Deteriorated joints between construction materials
- Improper pitch of flashing components at openings

A few specific examples related to window constructions follow:

- Gaps or holes in window perimeter sealants require that the sealant be removed and new sealant properly installed.
- Proper flashings should be installed around the window systems to help move moisture from the interior of the wall system to the exterior.
- The connection between window frames and mullions should be installed in accordance with the manufacturer recommendations and properly sealed to avoid gaps or openings in the window system.

FIGURE 6.39 Peeling paint underneath window opening.

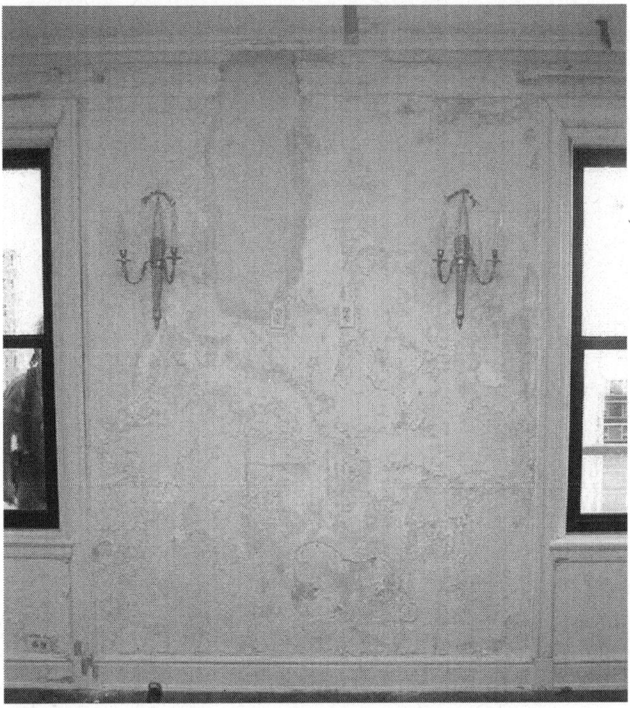

FIGURE 6.40 Peeling paint and plaster damage adjacent to window openings.

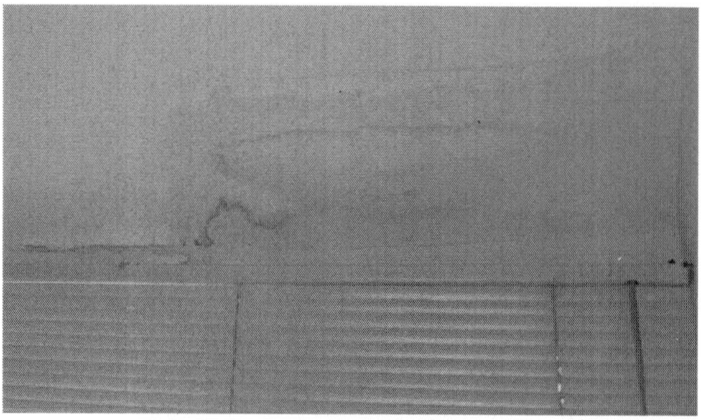

FIGURE 6.41 "Water stained" ceiling above a window opening.

FIGURE 6.42 (See color insert.) "Water staining," peeling paint, and mold growth at a garden-level bedroom.

FIGURE 6.43 "Spray rack" water testing equipment at window openings.

FIGURE 6.44 (See color insert.) Removal of interior drywall finish to observe water leakage during testing.

FIGURE 6.45 Water dripping on interior window sill during testing.

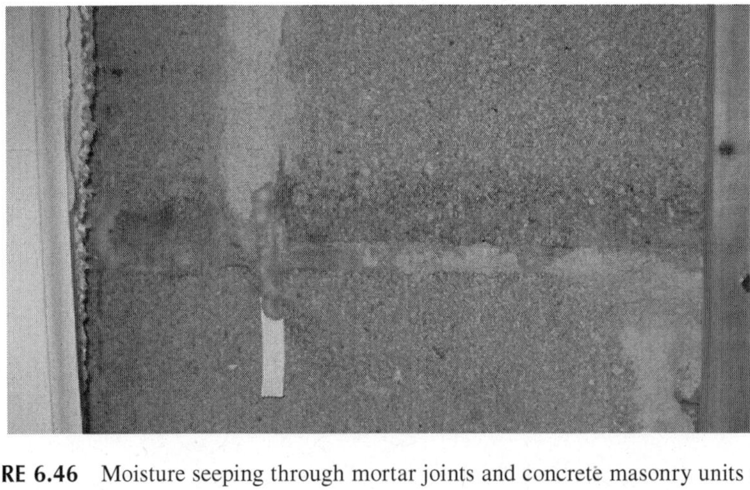

FIGURE 6.46 Moisture seeping through mortar joints and concrete masonry units during testing.

FIGURE 6.47 Moisture seeping through exterior wall system during testing.

FIGURE 6.48 Moisture seeping through exterior facade during testing.

FIGURE 6.49 Water seeping through exterior facade and roof during testing.

FIGURE 6.50 Exterior facade with failed mortar joints and improperly installed sealants.

6.9 ADDITIONAL CONSIDERATIONS FOR FORENSIC INSPECTIONS

6.9.1 Follow the Water

Forensic investigators should realize that molds grow on moisture films either in or on the surfaces of materials and not in room air. Following the pathway of water entry into building components is essential for understanding and controlling causes of mold growth in buildings. Moisture content readings (e.g., conductivity meters; see Section 6.4) can provide the forensic investigator with a practical method of finding niches where mold growth may occur. Chapter 7 of the Institute of Inspection Cleaning and Restoration Certification Standard 500 (IICRC, 2006) discusses the use of moisture meters and infrared cameras useful in inspections for dampness conditions. Chapter 5 in the AIHA (2008) *Recognition, Evaluation, and Control of Indoor Mold* and the ASTM (1994) publication on *Moisture Control in Buildings* provide investigators with background information on moisture dynamics and moisture control in both commercial and residential buildings.

6.9.2 Hidden Mold

Sections 6.4 and 6.5 of this chapter provide the forensic investigator with general procedures that can be used for finding hidden mold in building assemblies. The presence of hidden mold in building components is important because this condition may result in degraded indoor air quality (Miller et al., 2000; Morey et al., 2003). There is a growing consensus that hidden mold should be physically removed (AIHA, 2001; EPA, 2001; Health Canada, 2004; New York City Department of Health and Mental Hygiene, 2008). It is important, therefore, during destructive inspection to document both the location and extent (surface area in square meters) of mold growth on building materials. Currently, there is uncertainty with regard to how small of an area of hidden mold growth can be left unremediated without risk of degraded IAQ (AIHA, 2008, Section 17.6). However, if chronic moisture and dampness problems in a building structure are ignored then it is likely that a small area of mold growth will become more substantial as further biodeterioration occurs.

6.9.3 Sustainable Construction Materials and Biodeterioration

Many modern construction materials such as MWS and delignified paper products are highly susceptible to biodeterioration. Dutch researchers (Adan et al., 2005) have proposed a classification system for interior finishing materials, namely, "resistant," "fairly resistant," and "sensitive" to biodeterioration and mold growth. Resistant finishing products are to be used in potentially damp locations such as bathrooms and swimming pools. Sensitive materials are, according to Adan et al. (2005), to be used in dry areas and not in the building envelope. An important aspect of the forensic investigations described in this chapter is to provide recommendations for use of moisture-resistant construction materials in locations subject to periodic or chronic moisture incursion or dampness.

6.9.4 Green Architecture and New Aspects of Forensic Investigations

Mold growth in buildings is primarily associated with the presence of chronically damp conditions in locations containing moisture-sensitive materials. Some aspects of "green architecture" offer new challenges for forensic building investigators. Thus, the presence of roof gardens, living facades (vertical green gardens) on the building envelope, and living (garden) walls within the building provide potential niches for moisture problems and mold growth on sensitive construction materials. Henshell (2005a, 2005b) describes forensic investigations of leaky waterproofing membranes beneath roof gardens. The need for forensic investigations, as in the case of a leaky waterproofing membrane beneath a roof garden, can be mitigated by attention to proper design, careful installation, and continuous maintenance of each new green building component.

REFERENCES

Adan, O. C. G., M. M. Sanders, and R. A. Samson. 2005. Sustained fungal control through interior finish performance requirements, In *Bioaerosols, Fungi, Bacteria, Mycotoxins, and Human Health*, ed. E. Johanning 482–490. Albany, NY: Fungal Research Group Foundation.

American Industrial Hygiene Association (AIHA). 2001. *Report of Microbial Growth Task Force*. Fairfax, VA: American Industrial Hygiene Association.

American Industrial Hygiene Association (AIHA). 2005. *Field Guide for the Determination of Biological Contaminants in Environmental Samples*, 2nd ed., L.-L. Hung, J. D. Miller, and H. K. Dillon, eds. Fairfax, VA: American Industrial Hygiene Association.

American Industrial Hygiene Association (AIHA). 2008. *Recognition, Evaluation, and Control of Indoor Mold,* B. Prezant, D. Weekes, and J. D. Miller, eds. Fairfax, VA: American Industrial Hygiene Association.

American Society for Testing and Materials (ASTM). 1994. *Moisture Control in Buildings*, MNL 18. Philadelphia: American Society for Testing and Materials.

Arnow, P., J. Fink, D. Schlueter, J. Barboriak, G. Mallison, S. Said, S. Martin, G. Unger, G. Scanlon, and V. Kurup. 1978. Early detection of hypersensitivity pneumonitis in office workers. *American Journal of Medicine* 64: 236–242.

Blomquist, G. 1994. Chapter 14, The Book of Leviticus. In *Health Implications of Fungi in Indoor Environments*, R. Samson, B. Flannigan, M. Flannigan, A. Verhoeff, O. Adan, and E. Hoeskstra, eds. Amsterdam: Elsevier.

Environmental Protection Agency (EPA). 2001. *Mold Remediation in Schools and Commercial Buildings*, EPA 402-K-01-001.Washington, DC: EPA

Health Canada. 2004. *Fungal Contamination in Public Buildings: Health Effects and Investigation Methods*, H46-2/04-358E. Ottawa, CA: Health Canada.

Heller, R. M., T. Heller, and J. Sasson. 2003. Mold: "Tsara' At," Leviticus, and the history of a confusion. *Perspectives in Biology and Medicine* 46: 588–591.

Henshell, J. January 2005a. Waterproofing under green (garden) roofs, part 1 of 2. *RCI Interface*, 27–36.

Henshell, J. January 2005b. Waterproofing under green (garden) roofs, part 2 of 2. *RCI Interface*, 27–34.

Hodgson, M. J., P. R. Morey, J. S. Simon, T. D. Waters, and J. N. Fink. 1987. An outbreak of recurrent acute hypersensitivity pneumonitis in office workers. *American Journal of Epidemiology* 125: 631–638.

Institute of Inspection Cleaning and Restoration Certification (IICRC). 2006. *Standard and Reference Guide for Professional Water Damage Restoration*, S500. Vancouver, WA: Institute of Inspection Cleaning and Restoration Certification.

Institute of Medicine (IOM). 2004. *Damp Indoor Spaces and Health.* Washington, DC: National Academies Press.

Miller, J. D., P. D. Haisley, and J. H. Reinhardt. 2000. Air sampling results in relation to extent of fungal colonization of building materials in some water-damaged buildings. *Indoor Air* 10: 146–151.

Morey, P. R., M. J. Hodgson, W. G. Sorenson, G. J. Kulliman, W. W. Rhodes, and G. S. Visvesvara. 1984. Environmental studies in moldy office buildings: Biological agents, sources, and preventative measures. *Annals of the American Conference of Governmental Industrial Hygienists* 10: 21–35.

Morey, P. R., M. C. Hull, and M. Andrew. 2003. El Nino water leaks identify rooms with concealed mould growth and degraded indoor air quality. *International Biodeterioration & Biodegradation*, 52: 197–202.

New York City Department of Health and Mental Hygiene. 2008. *Guidelines on Assessment and Remediation of Mold in Indoor Environments.* New York: New York City Department of Health and Mental Hygiene.

Samson, R. A. 1985. Occurrence of moulds in modern living and working environments. *European Journal of Epidemiology* 1: 54–61.

7 Practices in Identifying, Remediating, and Reoccupancy When Mold Occurs

Gary R. Brown, PE, QEP, CMC

Determining an adequate definition of the cause for the presence of mold, or microbial contamination, is critical prior to considering mold remediation. There are a significant number of microbial contamination investigative techniques, but the story rarely, if ever, begins or ends with microbial contamination itself. The key question that a microbial investigator or consultant must ask is why the contamination is present and that question nearly always leads to discovery of some type of moisture or water intrusion condition, in the building environment.

This is true because moisture is nearly always the "limiting factor" in microbial growth, and high moisture conditions or repeated wetting can cause microbial growth to continue on an ongoing basis, or amplify, with ongoing release of mold spores. The ongoing release of spores is what causes microbial contamination, with attended nuisance complaints and health impacts.

As a general rule, other potential sources of indoor air problems should be subject to an initial review, and ruled out, before beginning a mold investigation. It is not unusual for basic ventilation problems, odor problems, or exhaust problems (from either appliances or exterior motor vehicles), to be initially reported as suspect mold problems. Most microbial consulting practitioners will

- Visit the project site
- Listen carefully to complaints and self-reported health symptoms
- Inspect HVAC systems, including accessible drip pans
- Measure moisture, carbon monoxide, and carbon dioxide levels
- Determine if there is any periodicity to the complaints

Periodicity, which refers to the timing or frequency of complaints or self-reported systems can lead a consulting practitioner to suspect or specific types of ventilation problems or exhaust problems. Four common types of exhaust and ventilation problems are

- Where ventilation system intakes are too close to loading docks, or passenger or truck loading areas, complaints will typically occur in early morning when vehicles are left idling and intake air draws in vehicle exhaust.
- Late morning complaints can be an indication of inadequate fresh air ventilation, as carbon dioxide builds up in occupied spaces. "Stalling" of ventilation systems sometimes occurs when ambient outdoor temperatures are similar to indoor "comfort zone" temperatures as many HVAC systems are set to work on "temperature demand." When there is little temperature demand on the HVAC system, little air moves and carbon dioxide builds up.
- Combination problems can also occur. Experienced indoor air quality practitioners frequently encounter situations where the design layout on the roofs of buildings with soffits has boiler or other mechanical equipment exhaust stacks located such that "downwash" causes the exhaust to be drawn into a nearby intake air plenum. "Smoke tests" conducted on a day with the proper wind direction can typically identify this situation.
- Finally, HVAC systems are sometimes set in a mode where they circulate unconditioned air at night (with no heating or air conditioning to minimize humidity). This can cause or exacerbate mold problems. In some areas of the country, nighttime air, during many days of the year, has relative humidity at 60 percent or above. Adding outdoor air with high humidity to the indoor environment to take advantage of cooler temperatures is bad HVAC practice where high humidity conditions prevail.

The typical next step in checking indoor air in a building is an initial visual inspection looking for evidence of mold, typically followed by obtaining humidity and building material moisture readings. Building material moisture readings should generally be 20 percent or lower, with the exception of concrete. Relative humidity readings should generally be 40 percent or lower. Mold will tend to amplify (or grow) where building moisture readings are above 20 percent or humidity is above 40 percent.

Table 7.1 contains a list of building materials where mold growth can commonly occur. Materials such as wood, many types of insulation, and gypsum drywall are commonly prone to mold growth when moisture content is high, and, more so, when repeated wetting events occur. Concrete, as a material, can be prone to surface mold growth, but many devices used to measure moisture in building materials will not work successfully on concrete because concrete, as a material, contains water as an

TABLE 7.1
Materials on Which Mold Can Grow

• Wallboard	• Insulation
• Fabric	• Soil
• Dust	• Food
• Paint	• Rugs
• Ducts	

ingredient in the concrete mixture. Other tools are available, to determine whether mold growth on concrete surfaces, or within concrete block cavities, is of concern.

The investigative tools available to microbial consultants are varied once an actual moisture or microbial contamination is found. Professional judgment is used to determine

- How extensive the investigation must be
- Whether "invasive" testing is warranted
- Whether the problem found can be managed through best management practices or requires one or more forms of remediation

The most important aspect of mold investigative work is to be sure of the source of any mold or moisture problem and to be sure that it is fully corrected before mold remediation work starts. Likewise, of key importance, is confirmation of the extent of any mold remediation needed and that the type of remediation chosen will be effective.

Mold investigation activities can be divided into three categories, as follows:

- Preliminary sampling—To see if there is a significant concern, and if there is any sign of microbial species of concern or microbial species concentrations are present.
- Investigative sampling—To determine the specific place(s) where microbial contaminants that may be present in occupied zone breathing spaces, on building materials, and/or in concealed spaces or HVAC systems.
- Evaluation sampling—To determine the extent of microbial contamination, leading to selection of the type of remediation work, which may be needed once microbial contaminants of concern are found in significant concentrations.

A matrix of sampling approaches that can be used for sampling can be found in Table 7.2.

When discussing types of sampling, it is important to understand the concept of *speciation*. Speciation refers to the ability to identify mold by species and not just by genus. This is important because when it comes to toxic or risk assessment molds, all too frequently, only certain species of a genus may be toxic. An example of this is *Stachybotrys. Stachybotrys chartarum* is a toxic mold, which has been widely publicized. Other species of *Stachybotrys* are not considered to have high toxicity, as compared to *Stachybotrys chartarum*. A mold consulting practitioner in most cases, needs to be aware of the genus and species of mold present to determine the degree of remediation needed. It also should be noted that some types of sampling and analysis will produce quantitative results, while others will only produce qualitative results.

Keeping these aspects in mind, an overview of the types of sampling typically conducted during mold investigation and evaluation work is as follows:

- Bulk sampling can be completed when a specimen of building material can be simply cut out and sent to the laboratory. An example would be gypsum drywall with mold growth on it, where a section of the wallboard is cut out

TABLE 7.2
Microbial Sampling Matrix

Sample Type	Speciate?	Detect Amplification?	Quantitative	Qualitative	Detect Airborne Exposure	Destructive	Other
Bulk	Yes	No	Only to media sampled	Yes	No	Yes	
Swab	Yes (unless preservative)	No	No	Yes	No	No	
Tape	Genus level only	No	No	Yes	No	No	
Culture Anderson N-6	Yes	Yes	Yes	Semi	Yes	No	Various agars available; report in CFUs
Spore trap[a]	Yes	Yes	Yes	Semi	Yes	No	Cannot do bacteria
Wall check	Yes	Yes, even if not amplified	Yes, only to air within wall	Semi	No	No	Filter to eliminate Exposure concerns

[a] *Identifies only amerospores: Aspergillus, Penicillium, Trichoderma.*

following appropriate health and safety procedures by the sampler, placed in a plastic bag, and sent to a qualified laboratory. Speciation is typically possible, but the sample results are quantitative only to the media sample. This sampling method does not detect airborne mold spore exposure.

- A swab sample involves removing a specimen of the actual mold on an accessible surface. The results are qualitative, and the method does not detect airborne exposure. Speciation is sometimes possible but not if a preservative is used.
- Tape sampling involves placing a tape on an accessible surface and removing a portion of the mold present as a specimen. This sampling method produces qualitative results, but results are to a genus level only.
- The spore trap method can be used to speciate and can determine if amplification is occurring, and it can be conducted to produce a quantitative result if the laboratory follows the appropriate analysis methodology. A spore trap method cannot be used, however, for bacteria. As this method uses an air sampling protocol, it can determine if "amplification" is occurring.
- A microbial culture sample can be obtained using an Anderson N-6 Air Sampling device with agar plate. This method produces a quantitative result and, as it is an air sampling technique, can detect amplification and provide speciation.
- A "wall check" sample can also be obtained from a sampling port placed in a wall. This method has the same characteristics as the Anderson microbial culture method, but the quantitation only applies to the air space in the wall, not the indoor air atmosphere. A filter can be used with the wall check method to eliminate any exposure concerns.

Mold, because it grows at variable rates in the environment, almost always in response to moisture conditions, is not an environmental toxicant or pollutant that is amenable to the application of strict numerical guidance levels or limits. At the current time, the practice of microbial investigation, evaluation, and consulting work involves comparison of sample results between areas where microbial impact is suspected and where it is not suspected. For microbial sampling and analytical results to be valid, an appropriate sampling plan needs to be formulated, which includes obtaining samples away from the suspect mold-impacted area and outdoors as well.

In some instances, where microbial amplification has continued for a long period of time, spores can be present in HVAC systems or throughout a facility or section of the facility in carpets, entry area mats, and so forth. In such instances, although the microbial growth may have occurred in a relatively limited area, the length of time over which spores proliferated and the resulting degree to which a more extensive facility or sectionwide cleanup needs to occur frequently can only be determined through air and surface sampling.

Additional investigative tools are also frequently used when mold growth is visually evident or water-damaged materials are found. These tools include

- Tomography, which is used to produce an infrared-like photo that illustrates, for example, where wetted insulation is present behind walls, not visible to the naked eye.

- Intrusive inspections, usually using a boroscope, can be used to see if mold is present in wall, ceiling, or floor cavities. A small diameter hole is drilled to facilitate a boroscope inspection.
- A specially trained mold dog can be used, which is particularly helpful in large buildings, where mold has become a persistent issue and concern, and other methods will be extensive or time consuming.

Once a mold consulting practitioner decides that mold remediation is indicated, consideration must be given to eliminating any moisture or water intrusion conditions that contributed to the mold growth and selecting the most appropriate type of remediation for the project.

When discussing mold remediation and how to approach it, several aspects of building construction and the characteristics of mold are important. These aspects are as follows:

- Small areas of microbial growth can contain large quantities of spores. The density of spores can be as high as 10,000 spores per square inch.
- Mold spores are frequently harder to clean up than asbestos fibers or lead dust. Although much of the technology for remediation used for addressing microbial concerns is also used for asbestos and lead paint projects, mold spores can sometimes take a number of days to "settle out," and they will keep growing in some instances and as fast as they are cleaned up, unless relative humidity is controlled.
- "Air scrubbing" using high-efficiency particulate air (HEPA) filters, along with HEPA vacuuming, may not be enough to bring a mold project to a successful conclusion. The sequences in staging of tear out, treatment, vacuuming, inspections, and air scrubbing as well as project critical barrier segmentation placement has to be carefully considered for all projects.

If significant concentrations of *toxic* or *risk assessment* microbial species are present, careful attention should be made to removing sensitive populations from the work area proximity. Most self-reported complaints and younger members of the population are considered more vulnerable to health impacts from spores. Protective measures implemented during mold remediation projects can include

- Implementation of critical barriers and negative air with HEPA filtration
- Use of independent professional oversight and inspection and verification sampling prior to reoccupancy
- Provision of temperature and humidity control to inhibit further microbial growth during remediation
- Complete removal and replacement of HVAC distribution systems where adequate remediation is not possible due to the presence of interior duct insulation or access limitations
- Appropriate use of microbicide to address any residual levels of spores, which may be present in the work area

TABLE 7.3
Mold Types of Concern

Genus	Species	Concerns and Symptoms	Microtoxins	Code[a]
Acremonium	—	Allergenic	—	—
Alternaria	—	—	A number of them	—
Aspergillus	*Flavus*	Allergic Bronchial Aspergillosis	Aflatoxin	R, C
	Fumigatus	Toxigenic, pathogenic	Many	R
	Nidulans	Carcinogenic, toxigenic	Yes	—
	Niger	Pathogenic	—	—
	Sydowii	Pathogenic	—	—
	Versicolor	—	Many	—
		—		
Cryptococcus	—	Old pigeon droppings; meningitis	—	—
Epicoccum	—	Human allergen; saprophyte	—	—
Fasarium	*Moniliformae*	Human allergen; saprophyte	T-2	R
Memnoniella	—	Similar to *Stachybotrys*	T-2	
Muncor	—	Opportunistic to leukemia		
Penicillium	*Aspergillus*	Eye, ear, lung, urinary tract, heart infections	Ochratoxin	C
Rhizopus	—	More than 200 species; opportunistic and systemic	Many; yes	—
Stachybotrys	*Chartarum*	MVOC chronic fatigue syndrome; lack of ventilation	T-2	R
Trichoderma	—	Pathogenic, sewage related; lung/liver infections	T-2	—
Ulocladium	—	Allergen	—	—

[a] *R, requires risk management; C, carcinogenic.*

Table 7.3 contains a listing of mold species considered to be *risk assessment* molds or those that are considered *toxic*. When considering remediation on a specific project, microbial consultants frequently refer to the *New York City Department of Health Mold Remediation Guidelines.*

For larger projects, preparation of specifications by a certified microbial consultant or professional engineer may be appropriate and necessary. Table 7.4 contains a typical specifications outline.

It is important that specifications and accompanying plans include information on any replacement building materials and HVAC systems. If microbial contamination has occurred because of underlying defects in design or construction of building or HVAC systems, all items needed to prevent the recurrence of a microbial contamination and moisture intrusion should be included in the plans and specifications package. Finally, all projects (even those not involving microbial contamination remediation) should give careful consideration to *protection of*

TABLE 7.4
Specifications Outline

	Page
1. DESCRIPTION OF WORK	
A. PART 1 – GENERAL	1-1
1. Description of Work	1-1
2. Scope of Work	1-3
B. PART 2 – MATERIALS	1-6
C. PART 3 – EXECUTION	1-7
1. Pre-Abatement Activities	1-7
2. Preparation of Work Areas	1-8
3. Removal Details	1-8
4. Final Cleaning and Clearance Procedures	1-8
5. Air Sampling Methodology	1-9
6. Post-Abatement Activities	1-9
2. TECHNICAL SPECIFICATIONS	
A. PART 1 – GENERAL	2-1
1. Overview	2-1
2. Definitions	2-2
3. Applicable Regulations and Codes	2-6
4. Submittals and Notices	2-7
5. Site Security	2-10
6. Emergency Planning	2-11
7. Personnel Protection and Safety	2-12
9. Training Requirements	2-21
B. PART 2 – PRODUCTS	2-24
1. Materials	2-24
2. Tools and Equipment	2-25
C. PART 3– EXECUTION	2-28
1. Pre-work Meeting	2-28
2. Preparation of Work Area	2-29
3. Isolation of the Work Area	2-31
4. Mold Removal Guidelines	2-35
5. Clean-up Procedures	2-40
6. Disposal Procedures	2-40
7. Transportation of Regulated Materials	2-42
8. Disposal Requirements	2-42
9. Environmental Sampling	2-43
10. Re-establishment of the Work Areas	2-43

materials, where there are semihumid or humid environments. This is important due to the following aspects:

- Building materials left in humid environments, and particularly those stored and not covered, have shown instances of mold growth, in a matter of weeks or months.
- "Stacks" of materials are hard to dry out, once wetted.
- Defects in building envelope design can typically take a year or two before there is evidence of mold growth after construction, where the mold growth is caused by cold weather insulation deficiency condensation conditions.
- Contractors and subcontractors working on projects where latent design defects become known later risk being drawn into building remediation or loss claims if they have not protected materials they furnished from microbial contamination impact.

With respect to mold projects, the New York City guidelines previously referenced, divide projects into categories as follows:

Level I—Small isolated areas that are 10 square feet or less (e.g., ceiling tiles, small areas on walls)
Level II—Midsize areas (10–30 square feet) (e.g., individual wallboard panels)
Level III—Large areas (30–100 square feet) (e.g., several wallboard panels)
Level IV—Extensive contamination (greater than 100 continuous square feet in an area)
Level V—Remediation of HVAC systems

A. Small, isolated area of contamination (less than 10 square feet) in the HVAC system
B. Areas of contamination (greater than 10 square feet) in the HVAC system

Once there is an indication that microbial remediation is substantially complete, water-damaged materials have been removed, and HEPA vacuuming and air scrubbing is completed, verification testing along with an inspection are typically undertaken. The inspection involves looking for any signs of remaining mold growth and may also include measurement of humidity and building material moisture levels. Assuming that the microbial consultant practitioner considers those elements satisfactory, verification sampling using either the spore trap or Anderson/culture sample method is then undertaken. Verification sampling frequently follows the same sampling approach as previously used during the investigative or evaluation stage and includes taking one or more samples away from the work area as well as an outside sample comparison. Successful verification sampling will typically demonstrate that there is no significant amplification still occurring in the work area and no more than single digits of toxic or risk assessment species, expressed as colony-forming units per cubic meter of air (CFU/m3).

The level and type of protective measures that are recommended to be implemented during remediation, as previously discussed, vary based on the nature and scale of the

TABLE 7.5
Protective Measures During Remediation: Summary of NYC DOH Levels

	Containment?	Who?	Protection	Other
		GENERAL		
I—<10 SF	No	MW – ok	R,G,E	Remove Susc. Indiv.
II—10–30 SF	Yes	MW – ok	R,G,E	Remove Susc. Indiv.
III—30–100 SF	Yes	TW w/O	R,G,E	Remove Susc. Indiv.
IV—>100 SF	Yes	TW w/O	R,G,E,S	Remove Susc. Indiv.
		HVAC		
VA—<10 SF	No	MW – ok	R,G,E	Remove Susc. Indiv.
VB—>10 SF	Yes	TW w/O	R,G,E,S	Remove Susc. Indiv.

Note: MW, maintenance workers; TW w/O, trained workers with oversight; R, respiratory protection; G, gloves; E, eye protection; S, Tyvek suits

project. Table 7.5 provides a summary of the recommended measures from the New York City guidelines. When a large-scale microbial impact remediation needs to be undertaken, the New York City guidelines recommend that the building owner, management, or employer notify occupants of the affected area(s) of mold presence. Where there are individuals with persistent health problems that appear to be related to bioaerosol exposure, physician referral and further evaluation of the exposure is recommended.

The results of verification testing, as well as all appropriate project documentation, including field logs, disposal records, air-testing results, contractor certificates of insurance, and project photographs should be included in a project report that also provides an opinion that the project area is suitable for reoccupancy. Unlike other forms of environmental cleanup (for example, asbestos and lead paint), mold growth or contamination can recur unless building owners and managers follow best management practices when maintaining their buildings. A list of suggested best management practices in buildings can be found in Table 7.6.

An aspect of property and building management that sometimes does not receive the attention it deserves is the need for tenant or building space mothballing. The term *mothballing* typically refers to a situation where building space has been vacated and there is no plan for near-term reuse. In such instances, building owners or property managers will seek to reduce building carrying costs by shutting off utilities and keeping the building in what they believe to be a "standby" mode. To prevent loss of building value, such buildings need ongoing attention to prevent mold and moisture conditions, particularly if relative humidity levels will no longer be controlled. To prevent microbial impact, additional aspects of mothballing above and beyond typical building "maintenance" should include

- Inspecting buildings weekly
- Maintaining basic electrical service or provide separate service, such that the building can be ventilated on dry days to prevent relative humidity buildup
- Maintaining the roof so as to prevent leakage

TABLE 7.6
Best Management Practices to Control Mold in Buildings

1. Respond immediately to any reports of water leakage or musty/unusual odors.
2. Use and encourage tenants to use only HEPA vacuums for cleaning to control dust to minimal levels. A highly effective program to remove dust from remediated areas is recommended.
3. Maintain the building roof so as to avoid any leakage.
4. Keep basements and crawl spaces dry; make all surface drainage flow away from building walls.
5. If the building is flat-roofed, maintain the roof in accordance with manufacturer's instructions. Re-roof prior to expiration of roofing warranty.
6. Maintain windows and doors to minimize water intrusion into porous building materials and furnishings. Inspect interior perimeters of all windows and doors twice per year.
7. Maintain exterior walls to eliminate rainwater infiltration. Inspect flashings, caulked joints, and exterior door/window edges twice per year.
8. Be aware that water damaged insulation and/or drywall should never be allowed to remain in any building.
9. Carefully inspect all areas where building materials were subject to water damage due to flooding, pipe leakage, overflows, or due to water damage or condensation from any other source. Inspect behind walls and ceilings.
10. Remove all water-damaged materials within 36 hours.
11. Encourage tenants or other occupants to report water damage and/or suspect mold growth immediately.
12. Get prompt professional assistance in the event of repeated indoor air quality complaints, self-reported mold-related health effects, and observation of significant water damage before undertaking mold remediation.

- Maintaining sumps, pumping systems, and stormwater and sanitary drainage systems, such that backups do not occur
- Removing any water-damaged materials including carpets and insulation within 36 hours
- Installing "key component" alarm sensors with backup to give an "early warning" alarm to prevent inundation of building materials

Further, it is important to understand the characteristics of buildings that cannot be successfully remediated, which can occur for a number of reasons. These include

- Buildings where an interior, insulated duct system is present and in-place remediation is not possible without the opening of walls and ceilings to remove and replace the HVAC distribution or return system
- Buildings where there have been repeated inundation events of stormwater or sewage, and it is not practical to remove any and all water-damaged materials
- Buildings, including the aforementioned and others, where the cost of remediation and building materials replacement exceeds the value of the building and construction of a new building or section of a building would provide a more cost-effective solution (where the costs of remediation and restoration begin to increase to 40 to 60 percent of a building's value, replacement is usually the chosen option)

Last, with respect to water damage claims and insurance issues, the path to successful remediation and reimbursement is sometimes unfortunately clouded. Where high wind, tornado, hurricane, and other storm damage occurs, which opens up the indoor environment to outside air, where repair will not be immediate, and heat and humidity may not be controlled for some time, the stage is set for mold growth. Yet, in many instances, mold growth is not covered by property insurance policies.

Proper mitigation of water damage and inundation events by a mold professional practitioner means removing all water-damaged materials within 36 hours taking prompt steps to eliminate further water intrusion, and promptly controlling temperature and humidity.

There is no "bright line" between condensation, inundation, and other types of water damage prior to the beginning of mold growth, but prompt action can make a difference as to whether a building or a section of the building is a total loss, or is a candidate for successful restoration. Although insurance carriers will generally not honor standalone mold claims, where microbial growth has greatly increased the cost of remediation, carriers and their preferred restoration service providers have responded to mold remediation claims where the extent of mold remediation work needed greatly increased after a sudden and accidental water intrusion or inundation event was not properly attended to while the building was under their control.

The science of mold investigation and remediation continues to evolve based on advances in medical science. Mycotoxins and volatile organic compounds from a wide variety of fungi are identified as potentially leading to *exacerbation* of the immunologic (allergic) reactions, causing health affects, or causing infections. For this reason, practitioners in microbial consulting should stay abreast of microbial science developments. Similarly, people or companies using microbicide to control mold growth on more than a localized, incidental, or infrequent basis are cautioned against chemical overuse. Overuse of microbicides should be avoided, where mold growth problems are repeated and persistent.

8 Analysis of Microscopic Contaminants in Sick Building Investigations

James Millette, PhD
Barb Epstien, MPH, CIH
Elliott Horner, PhD

CONTENTS

8.1 INTRODUCTION

Microscopy is a useful tool to identify mold found during sick building investigations. Microscopy can also play an important role when the cause of a problem turns out not to be mold and the investigator must consider the other particulates present at the site.

There are several types of sick building situations where microscopy may play an important part. The first situation involves sampling of obvious mold colonization on surfaces. Another situation is based only on the complaints of the occupants where other indicators such as musty odors or obvious mold conditions do not immediately appear to exist. Here the investigator must rely on several investigative tools, including the collection of surface dust samples or the collection of airborne particulate samples, in an attempt to eliminate some potential causes of the problem and narrow the possibilities. Still another situation occurs where staining or discoloration of building surfaces cause concerns that may be related to occupant health problems. This chapter will discuss some of the sampling and microscopical analysis methods used during the investigations of sick building complaints.

8.2 GENERAL SAMPLING CONSIDERATIONS FOR PARTICULATES

Surface dust particle collection techniques vary according to situation. From the microscopist point of view, the preferred collection media for most sick building investigations is a combination of a tape lift and a clean-room wipe of the surface material. Other techniques include scraping, brushing, Post-It notes, scoop and bag techniques, and vacuuming with either a microvac (air cassette) or large-scale vacuuming (Millette and Few, 2001). For situations where the concern is a stain or darkening agent, a combination of tape lift and cotton ball (or cloth wipe) collection media as described in the ASTM Standard Practice D6602 (American Society for Testing and Materials [ASTM], 2003) has been found to be most useful for the analyst to examine. When the area to be sampled is a porous material such as fabric or carpet, the microvacuum sampler is useful. It is also valuable when the concern is about exposure to particles that can be readily entrained into the air during dust disturbance. The microvacuum sampler consists of an air cassette with a filter inside that has been modified with a nozzle or collection head. It is attached to either a battery-powered personal pump or a mains-powered area high-volume pump. Cotton swabs that are used in microbial sampling can be analyzed for other particulates but the limited amount of material on the swab makes it difficult to use with multiple microscope techniques that are necessary to identify all the particle types present. An advantage of tape lifts is that the spatial arrangement of fungal particles is better preserved than with swab or wipe samples. The arrangement of particles can aid in distinguishing between settled particles in dust and particles formed on the surface by colonization, for example, particularly with *Cladosporium*.

Some building investigations require sampling of actual building materials such as wallboard, ceiling tiles, or duct lining. This type of sampling involves removing the material in a manner that will maintain, as much as possible, the layer structure and integrity of layered samples. For wall systems this can be accomplished with a polycarbonate tube used like a cork borer or by carefully cutting through the product down to the substrate, wrapping in paper and cushioning the sample so that the product cannot be crushed or crumbled.

Because objectives for each building investigation and dust and particulate characteristics differ from site to site, there is no single sampling method of choice for all situations. In most cases it is best to collect several samples of dust particles with more than one sampling medium and let the laboratory choose which is best for analysis.

Samples should be properly documented with field identification information placed on the sample bag or container. Chain-of-custody forms should accompany the samples to the laboratory.

8.3 SAMPLING FOR MOLDS

The following is a brief description of some of the methods used to sample mold for microscopical examination. These include air, source, and surface sampling and analysis. A full discussion of sampling strategies, including selection of the type(s) and number of samples, is beyond the scope of this chapter and may be found in other chapters contained in this book. Keep in mind, however, that there is no one-size-fits-all sampling method for mold and also that sampling is only one of several tools that should be incorporated into a sick building investigation (e.g., walk-through visual inspection and review of occupant concerns, general building performance, environmental conditions, etc.). Each method has inherent strengths and limitations. Consequently, using a combination of sampling techniques will help the investigator to obtain appropriate and useful results for interpretation, and all require prior development of an appropriate sampling strategy. As with any environmental investigation, limited amounts of data usually support only limited conclusions. Thus, data from only a few samples can justify further investigation but would not firmly support the conclusion that the building should be declared sick.

8.3.1 Air Sampling

Generally, investigators collect air samples to assess occupant exposure to bioaerosols. Common air samplers include culture-based, multiple-hole impaction devices, which collect airborne fungal propagules for growth on nutrient agar media, and nonculture-based spore traps, which impact the collected particles onto a coated adhesive surface rather than an agar media. In sick building investigations these typically are collected as area samples with individual samples usually collected over a relatively short duration (e.g., up to several minutes). In heavily contaminated work environments or where longer duration or personal sampling may be desired, air sampling with a filter cassette may also be employed. However, potential limitations of this method include high detection limits (if analyzed by direct microscopy) and desiccation, which can result in loss of culturability of some fungal elements (American Industrial Hygiene Association [AIHA], 2008).

Many of the basic principles of airborne particulate sampling are applicable to bioaerosol sampling (e.g., volumetric air sampling using calibrated air sampling instruments, collection efficiency). It is well established that "settle plates" (also known as gravity sampling) do not provide representative or reliable bioaerosol samples (Macher, 1999).

8.3.2 Source Sampling

When feasible to do so, it may be useful to obtain a portion of a suspect building material, for example, one that exhibits discoloration or is visibly water damaged.

Common examples are pieces of building construction, finishing, or furnishing materials, and building system components (e.g., duct lining, insulation). Bulk samples may also provide some utility even in the absence of obvious visible indications of mold growth but where the investigator has some knowledge of the material having been affected by moisture damage or dampness. Bulk sample collection should be performed in a manner that will avoid or minimize cross-contamination of the sample or the surrounding environment. The collected sample typically is placed in a clean, sealable plastic bag or rigid container (if needed to maintain the sample's structural integrity). When preparing bulk samples for transport to the laboratory the material's relative moisture content should be taken into account. If the material is damp, consider placing a packet of desiccant in a paper bag into the sample container (AIHA, 2008; Martyny et al., 1999). It may also be advisable to enclose a colorimetric humidity indicator strip, which provides a means of monitoring humidity levels inside containers.

Bulk material samples may be analyzed by direct microscopy, which can provide confirmation of the presence of mold colonization (growth) versus the mere presence or deposition of mold spores; typically this is a qualitative or semiquantitative analysis. The material also may be cultured on various agar media to allow for enumeration and identification (to species level where possible) of mold colonies present. It is useful to collect source samples both from moisture problem areas and from background, or unaffected, areas as reference controls (Morey et al., 2000).

8.3.3 Surface Sampling

Surface sampling is favored over bulk sampling when a less destructive technique is needed or desired. Collection of surface samples for analysis of molds usually is conducted by either (a) vacuuming of settled dusts, (b) tape lifts, or (c) wiping a surface with a swab.

8.3.3.1 Dust Sampling

Settled dust can be collected from either hard, smooth surfaces (e.g., metal or wood) or from soft, fleecy surfaces (e.g., upholstery). Dust samples can be collected using the microvac technique (filter cassette attached to a vacuum pump) or with various adaptors fitted to the nozzle of a standard vacuum cleaner. Surfaces being sampled should be dry prior to sampling.

It may be advisable to use a template to define the surface area to be sampled. A minimum dust loading is typically required, which may affect the size of the sample area. This in turn may vary depending on the nature and condition of the surface being sampled. Depending on the sampling strategy for a given investigation, individual samples may be collected and analyzed separately, or multiple samples from the same area or from similar surface types may be composited either in the field or in the lab (American Conference of Governmental Industrial Hygienists [ACGIH], 1999; AIHA, 2008). The amount of dust in a sample should be sufficient (e.g., at least about a teaspoon's worth) to allow for adequate processing in the lab and meaningful detection limits.

8.3.3.2 Tape Lifts

Tape lift samples are easy to collect and are a useful aid in evaluating suspect surfaces (e.g., discolorations observed subsequent to moisture damage). Analysis by direct microscopic examination provides semiquantitative assessment that enables the investigator to distinguish between a normal accumulation or deposition of common spore types, atypical accumulations of spore types that might be associated with adjacent or nearby mold growth, and the presence of confirmed fungal growth on a surface (these scenarios correspond with the descriptions of Conditions 1, 2, and 3, respectively, as defined by the Institute of Inspection, Cleaning and Restoration Certification [2008]). Microscopic examination of tape lifts may also help differentiate between microbial growth and accumulation of other, macroscopically similar particles and fibers (Edwards, 1951).

Tape lifts are collected by lightly pressing a strip of clear adhesive tape onto a surface (hold the tape by the edges only) and then placing the exposed tape onto a clear glass slide. Immediately prior to sampling, the investigator should remove and discard the leading couple of inches of tape to avoid contamination from previous use or storage. The sample should be about the size of a thumbprint. A sufficient sample will appear as a light deposit; too heavy a deposit or a lot of opaque debris (e.g., paint chips) will make the microscopic examination difficult or impossible. In this case, the investigator should repeat the sampling to obtain a light deposit from the area of concern. The choice of sampling location also affects the value of the resultant analytical data (AIHA, 2008; Martyny et al., 1999).

8.3.3.3 Swab/Wipe

Swab and other wipe samples are usually collected from hard, nonporous surfaces, using any of a number of commercially available swab kits. Swab samples are collected by wiping a surface with a moistened swab (for dry surfaces) or with a dry swab (for moist surfaces). Wiping typically is done using a rolling motion on a premeasured area. It is advisable to handle swab samples aseptically if the samples are to be analyzed by culture (AIHA, 2008; Martyny et al., 1999).

8.3.4 Images of Common Molds

Although it is not the purpose of this chapter to provide a detailed description of the analysis procedures for the identification of mold, light microscopy images are shown in Figures 8.1 to 8.6. These images provide some of the characteristics that are used in the identification of some common fungi. Three magnifications—low (approximately 50 times), medium (approximately 400 times), and high (approximately 1000 times)—may be used during mold identification. Figure 8.1 is a medium magnification image of *Alternaria alternata*, showing short chains of conidia (spores). *Alternaria* is commonly encountered in air, both indoors and outside. It colonizes a broad array of organic materials and is a common leaf surface inhabitant. Figure 8.2 shows a low magnification image of a piece of wood veneer colonized with *Penicillium brevicompactum. Penicillium brevicompactum* occurs widely in nature on a variety of substrates, including damp or water-damaged building materials. Figure 8.3 shows a low magnification image of an *Alternaria alternata* colony on agar. Note the

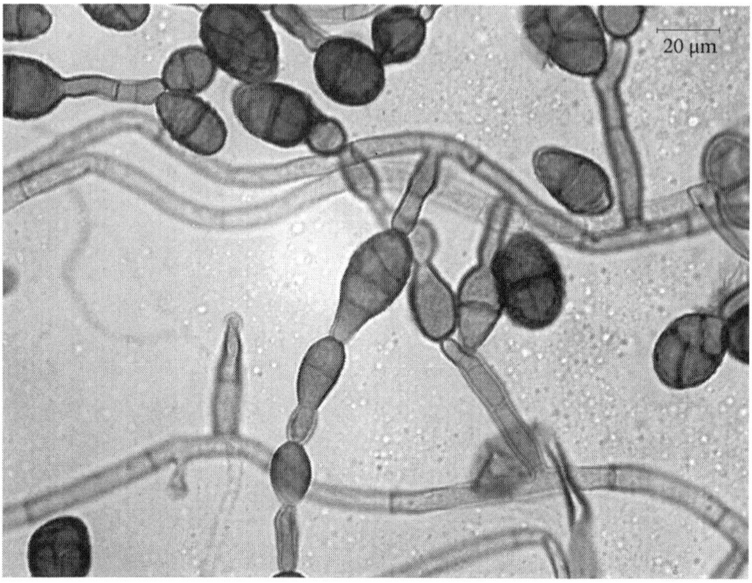

FIGURE 8.1 *Alternaria alternata*, showing short chains of conidia (spores).

FIGURE 8.2 Wood veneer colonized with *Penicillium brevicompactum*.

FIGURE 8.3 (See color insert.) *Alternaria alternata* colony on agar.

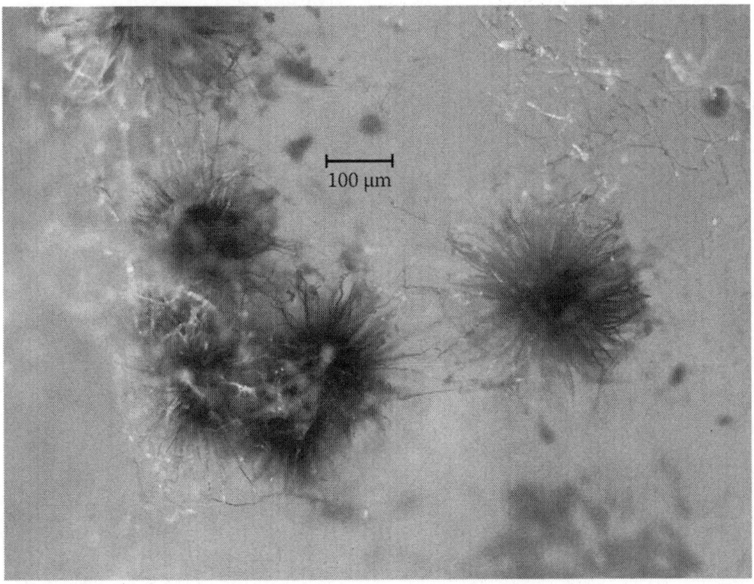

FIGURE 8.4 *Chaetomium globosum* fruiting structures (ascomata).

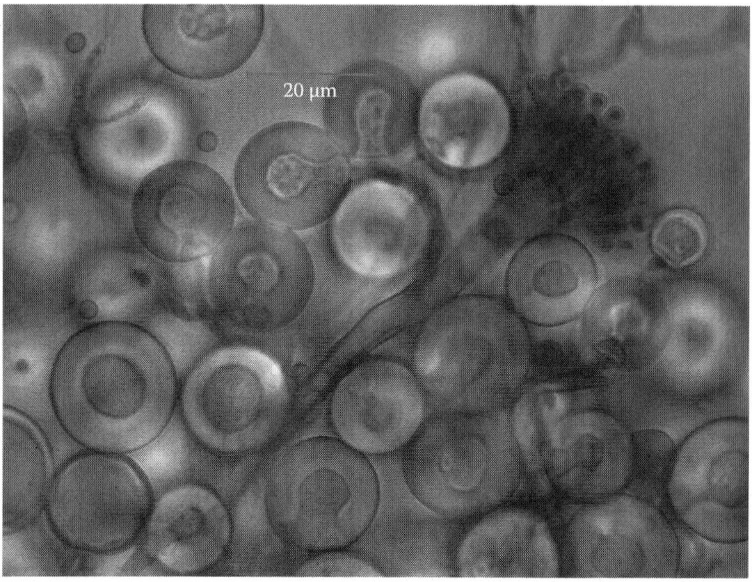

FIGURE 8.5 (See color insert.) Hülle cells and a conidiophre of *Emericella (Aspergillus) nidulans*.

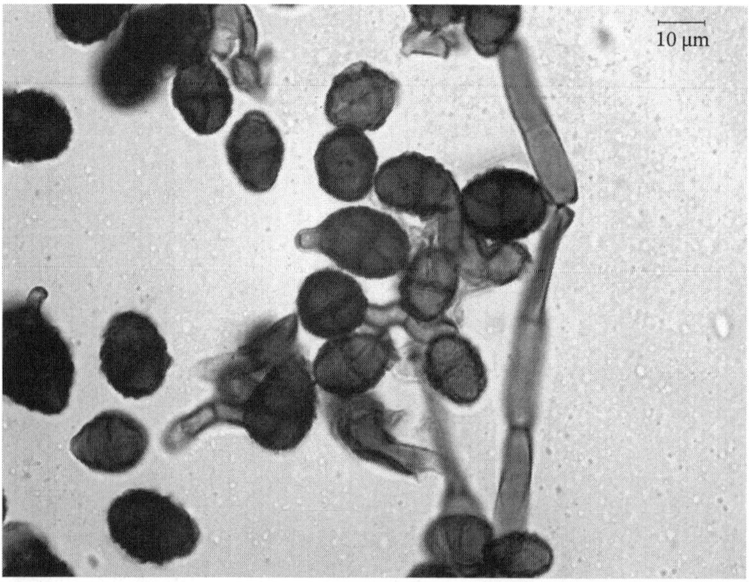

FIGURE 8.6 Spores of *Ulocladium chartarum*.

short chains of conidia. Many of the plant pathogenic species of *Alternaria* have conidial chains that are much longer. Figure 8.4 shows a low magnification image of *Chaetomium globosum* fruiting structures (ascomata). Note the spiral perithecial "hairs." *Chaetomium* is a very efficient degrader of cellulose and a cause of "soft rot" in chronically wet wood (e.g., wood water tanks). It also is an indicator species of water damage on gypsum wallboard. Figure 8.5 shows a high magnification of Hülle cells and a conidiophore of *Emericella (Aspergillus) nidulans*. Hülle cells are distinctive with the thick, refractive cell walls. Clusters of Hülle cells comprise the structure of the fruiting body of *Emericella*. Figure 8.6 shows a medium magnification image of spores of *Ulocladium chartarum*. Note the general similarity to the spore structure of *Alternaria*. *Ulocladium chartarum* is an aggressive degrader of paper (or *charta* in Latin) and thus is often found in water-damaged buildings. More detailed information about the analysis procedures for identification of mold may be found in other chapters of this book and can also be found in several references (Farr et al., 1989; Hawksworth et al., 1995).

8.4 MICROSCOPY USED IN SICK BUILDING INVESTIGATIONS

The microscopy portion of an initial building dust investigation usually begins with an examination of the particles using a low-magnification, reflected light (stereobinocular) microscope followed by polarized light microscopy (PLM). In some situations when dealing with the investigation of unknown dust particles, the PLM analysis is sufficient to answer a question about the presence of a specific contaminant. In other situations, other microscopy techniques such as electron microscopy or infrared microscopy may be needed to complete the analysis. The low-magnification microscopy and PLM are used to sort out the various particle types present and to determine the approximate relative percentages by volume of the different components. Common residential and building dusts contain biological substances including fungal spores (Figure 8.7), pollen grains (Figure 8.8), skin cells (Figure 8.9), plant fragments (Figure 8.10), and insect parts (Figure 8.11). Dusts also may contain combustion products including char soot (Figure 8.12), building materials such as glass fibers (Figure 8.13), drywall debris (Figure 8.14), and rust flakes (Figure 8.15). Both man-made and natural fibers are found in many indoor dusts. These include cotton fibers (Figure 8.16), synthetic fibers (Figure 8.17), and hair (Figure 8.18). Indoor dust particles can be classified by PLM examination using approximately 20 particle classes (Millette et al., 2003; Turner et al., 2005). The McCrone *Particle Atlas* is the standard reference for microscopical particle classification (McCrone and Delly, 1972). Scanning and transmission electron microscope (SEM and TEM, respectively) analyses are used in conjunction with energy dispersive x-ray spectroscopy (EDS) to identify inorganic particles as well as to characterize the "fine" (small) size fraction that includes particles such as aciniform carbon soot. Raman microspectroscopy and infrared microscopy, using Fourier transform infrared microspectrophotometry (micro-FTIR), are very useful when identifying fragments of organic building materials such as plastics and polymers as well as pharmaceutical and pesticide carriers.

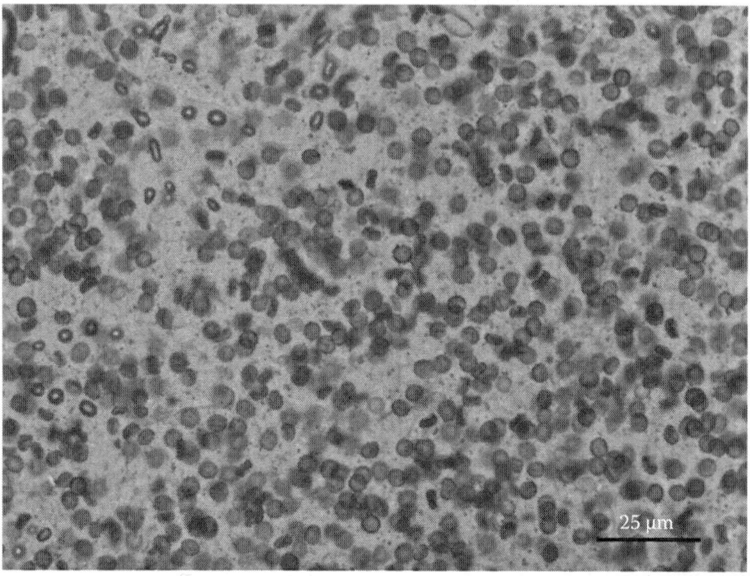

FIGURE 8.7 Light microscope image of fungal spores.

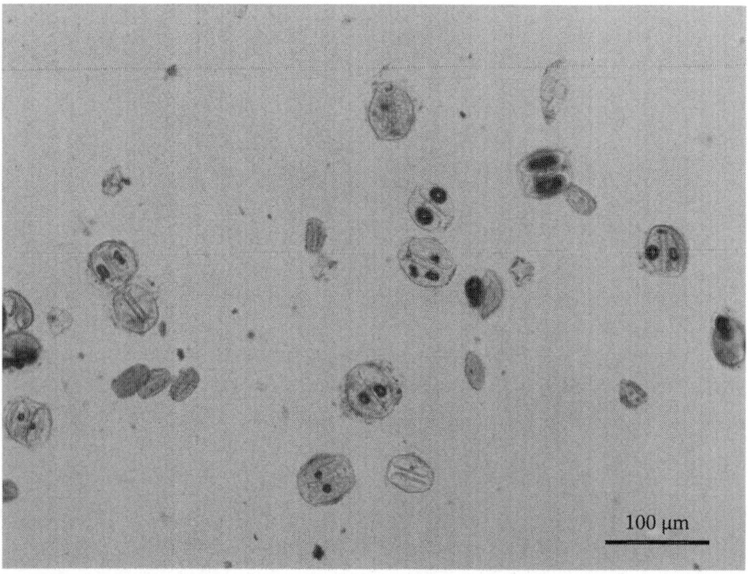

FIGURE 8.8 Light microscope image of pollen grains.

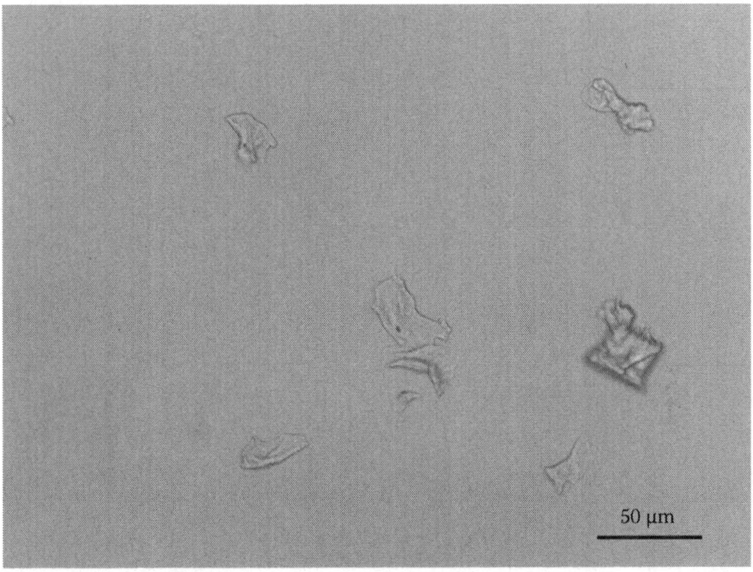

FIGURE 8.9 Light microscope image of skin cells.

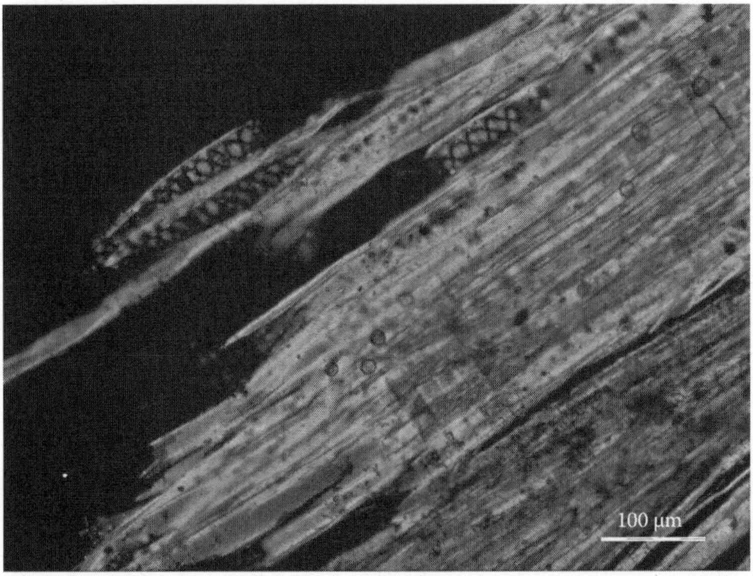

FIGURE 8.10 Light microscope image of plant fragments.

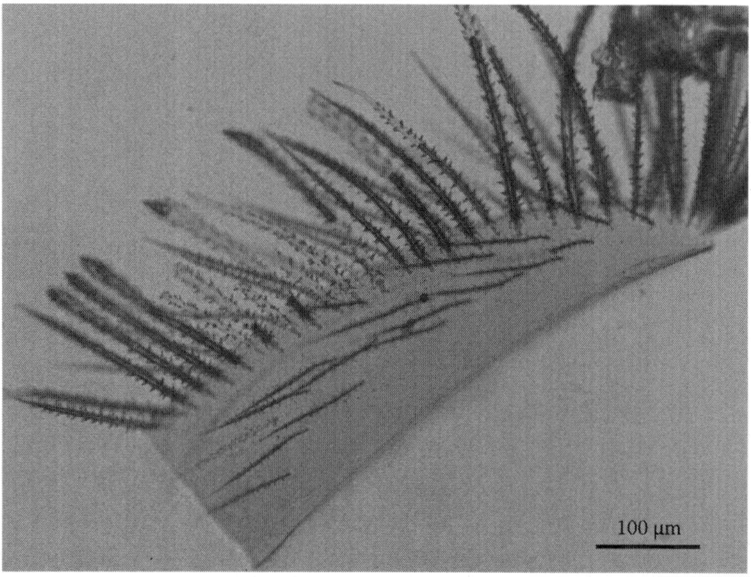

FIGURE 8.11 Light microscope image of insect parts.

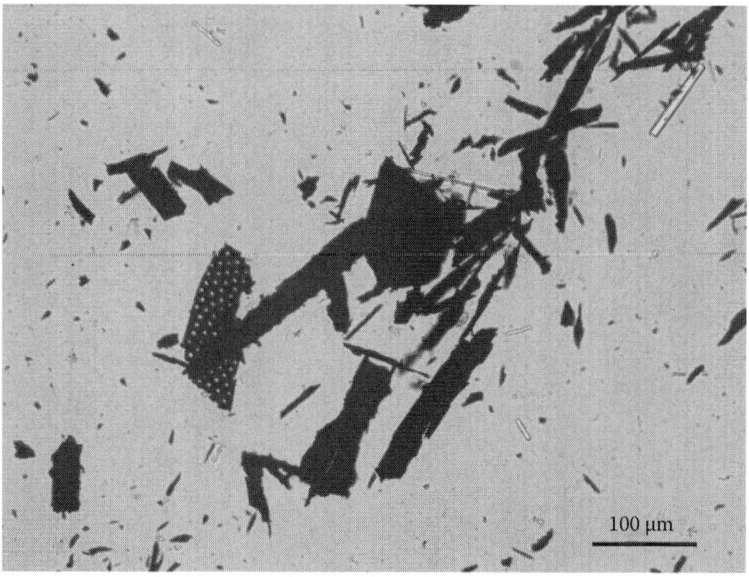

FIGURE 8.12 Light microscope image of char soot.

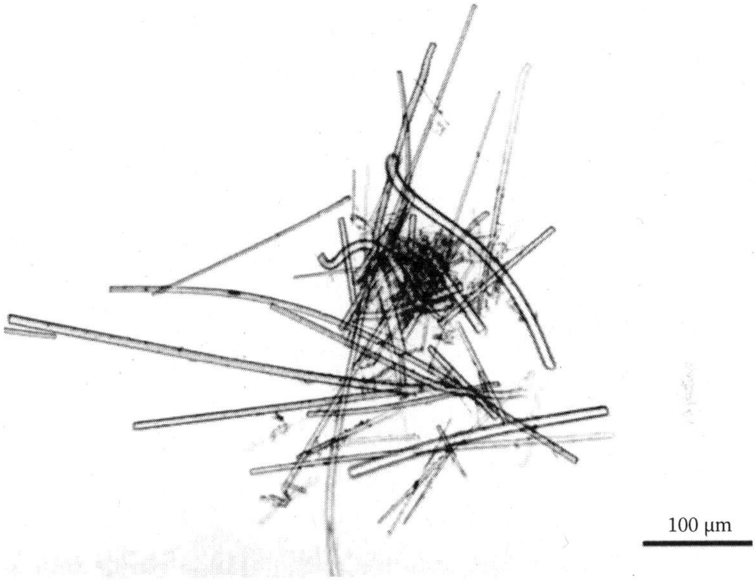

FIGURE 8.13 Light microscope image of glass fibers.

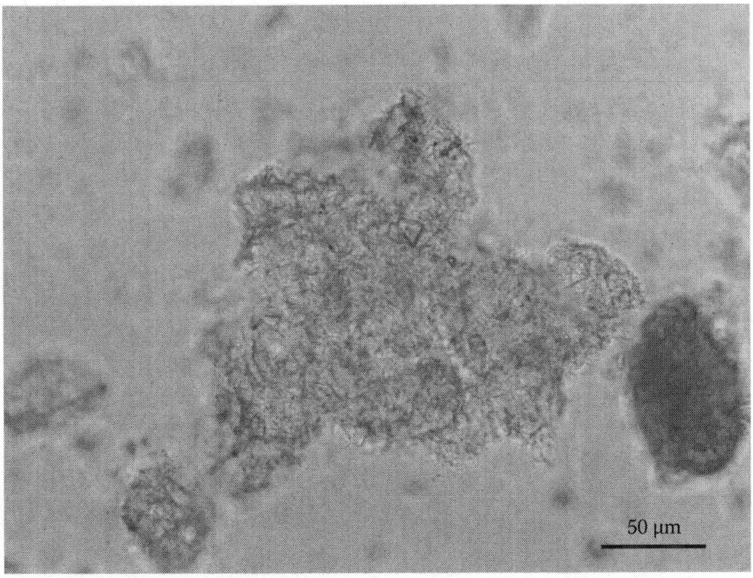

FIGURE 8.14 Light microscope image of drywall debris.

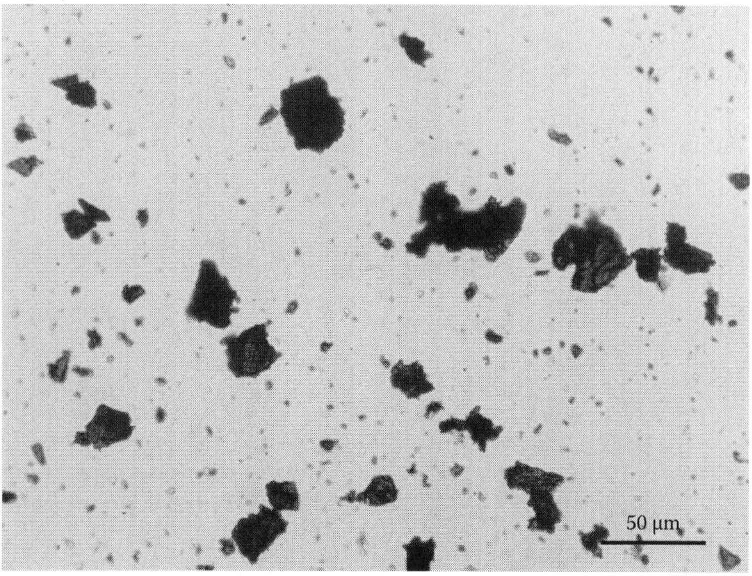

FIGURE 8.15 (See color insert.) Light microscope image of rust flakes.

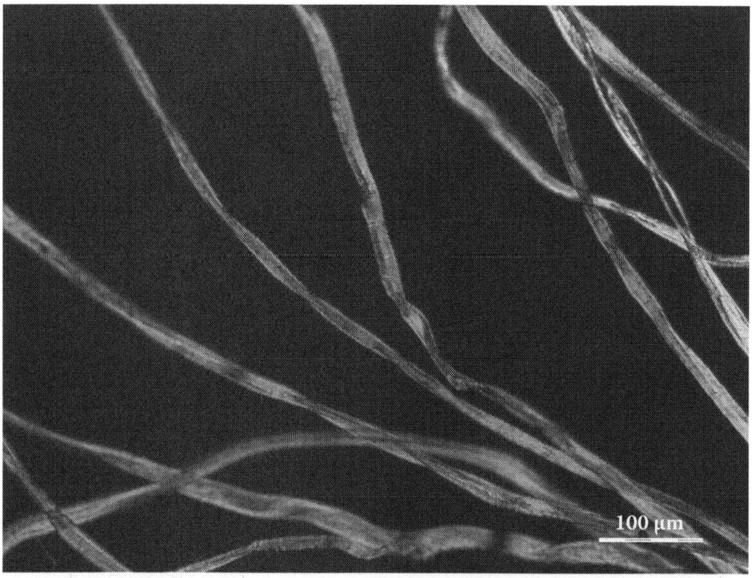

FIGURE 8.16 (See color insert.) Light microscope image of cotton fibers.

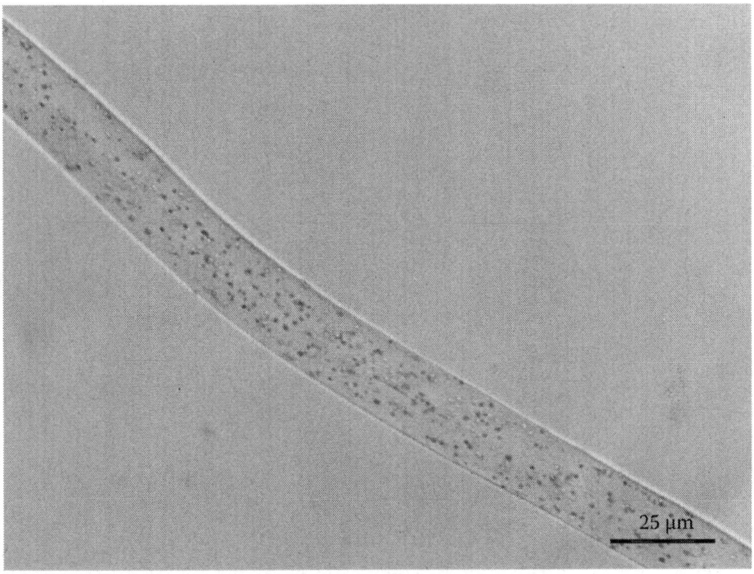

FIGURE 8.17 Light microscope image of synthetic fibers.

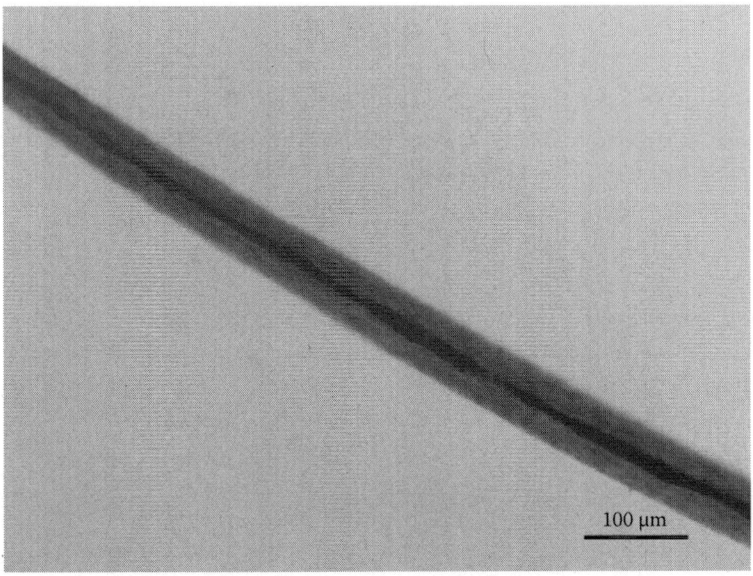

FIGURE 8.18 (See color insert.) Light microscope image of hair.

8.5 SURFACE DARKENING PARTICULATES

A number of materials (agents) when in contact with a surface will make the surface look dark. Sometimes an occupant of a building will assume that dark looking patches or stains are mold. *Ghosting* is a term used in some indoor situations (Millette, Gerber, et al., 2007). Among the common causes of undesired darkening of a surface are mold, soot, and dirt (Millette, 2001; Millette et al., 1994). Microscopy can be used to examine darkening agent(s) and quickly determine whether mold, soot, dirt, or a combination of the three is present. In the investigation of stains or darkening agents, the American Society for Testing and Materials Standard Practice 6602 (ASTM, 2003) is a standard procedure that serves as an excellent basis for microscopy investigations where darkening agents are involved. This ASTM standard was designed primarily for the determination of carbon black among soot particles and other dark particles but provides the framework for the microscopy studies necessary to determine the identity of all particles and possible sources of surface contamination. Polarized light microscopy is used first to determine general percentages of the types of particles present. If mold is the major component of the sample, additional light microscope analysis may be done to determine the specific type of mold present. If soot is found to be the major component of the sample, transmission electron microscopy is used to confirm the aciniform nature of the soot particles and characterize the soot particles so that a possible source may be determined.

8.6 CASE STUDIES AND NOTES FROM SICK BUILDING INVESTIGATIONS INVOLVING MICROSCOPY

Applications of microscopic techniques as an approach to understanding issues of exposure intensity and source identification were provided in an article about using dust analysis as a metric for exposure evaluation (Lioy et al., 2002). Other examples of environmental forensic microscopy investigations have been previously published and are summarized here for convenience (Millette et al., 2008; Millette and Brown, 2007).

8.6.1 Office Building Air Duct

In a typical case involving a sick building investigation, black and white particles were found in a governmental office building several days in a row on a desk located under an air system duct grate in the late fall. Light and scanning electron microscopy showed that the particles were black clusters of fungal spores and white fragments of galvanized metal corrosion product. The change in season had apparently changed the microenvironment inside the duct from one that was humid to one that was dry. The humid environment encouraged the growth of mold and the corrosion of the metal duct.

8.6.2 Office and School Building Glass Fibers

Exposures to glass fibers have been associated with some sick building syndrome symptoms. Glass fibers (also referred to as man-made vitreous fibers or machine-made vitreous fibers [MMVF]) may cause eye irritation, sore throat, sinus congestion and headaches, and rashes (contact dermatitis). Insulation products and ceiling tiles are potential sources of glass fiber exposure. Light and electron microscopy has been used to measure the amount of glass fibers per unit area of surface dust. Microscopy can also differentiate between the different types of glass fibers including slag wool and rock wool (Brown et al., 2007).

8.6.3 A Spot Called Ralph

In a courthouse in South Carolina a mysterious stain appeared in the new carpet. Employees even gave the spot a name—"Ralph." At first it was the size of a half dollar but it grew after cleaning to about 2 square feet. Environmental mold specialists initially tested the stain in the carpet and determined that it was not caused by mold. A section of the stain was cut from the carpet and delivered to the forensic microscopy laboratory for inspection. Analysis by light and scanning electron microscopy showed that the carpet contained a variety of particles typical of the particles often found in office dusts. A sticky substance was also found on the carpet fibers. FTIR analysis of the sticky material showed that it was consistent with corn syrup. It is apparent that someone spilled a soft drink on the carpet, and the stain was caused by office dust particles adhering to the sticky drink residue. Efforts to clean the stain removed the dark office dust particles but did not completely remove the sticky drink residue. In fact, the cleaning efforts spread the sticky residue, which then collected more office dirt over time and therefore appeared to grow in size. Not surprisingly, the newspaper reporter who interviewed the laboratory after the findings were made public was disappointed that the stain called Ralph was not something exotic but caused by a spilled soft drink.

8.6.4 Ghosting

One case concerned the investigation of "ghosting" also known as "black soot deposition." In this case microscopic examination of the black material by TEM and x-ray elemental analysis showed that the black material was soot particles consistent with carbon soot from paraffin burning, but it was established that candles were not used in the residence. Further investigation showed that the cause of the ghosting problem was soot from paraffin logs that had been used in the fireplace. A backdraft was apparently carrying the soot particles from the interior chimney walls throughout the house. Soot is the product of low-temperature combustion where there is insufficient oxygen for a complete burn. Soot can come in several forms (Goldberg, 1985; Medalia and Rivin, 1982; Millette and Few, 1998; Millette, Turner, et al., 2007). Soot from wood-burning fireplaces, the burning of leaves, and forest fires contains a considerable amount of "char." Char is the term for the microscopic particles of soot that have some remnant of the original material that was burned. A small portion of a charred leaf stem is an example of char soot. Spherical particles of soot called aciniform soot can be generated during oil burning (Stevenson, 1982). Very small soot particles that attach together and look like a bunch of grapes are called aciniform soot. Aciniform soot can be generated during the burning of most materials.

Sources of aciniform soot include: candles, oil lamps, propane grills, fireplaces, diesel exhaust, and the burning of organic products such as oil, gas, and chemicals. Carbon black is a commercial form of aciniform soot used in making tires, inks, and pigments for paint.

8.7 DISCUSSION AND CONCLUSIONS

Paul Lioy (2006) points out that our indoor environment contains many types of organic and inorganic mixtures that can be found simultaneously in the air or deposited on surfaces. Microscopy in sick building investigations allows an investigator to characterize the particles in a sample in ways that are different from the results of instrumental analyses. These characterizations augment the information determined by other tests and allow the determination of contaminant sources. With the information obtained through the environmental forensic microscopy study of the particle types, the investigator can more fully assess an individual's exposure to a particular contaminant.

Microscopy can be very useful in the initial stages of a sick building investigation effort where it is used to eliminate some concerns and to narrow the field of possible sources of the problem. It is important to remember that environmental forensic microscopy sampling and analysis is only one of several tools that should be incorporated into a building investigation. Other procedures such as a walk-through visual inspection, review of occupant concerns, general building performance, and environmental conditions should be considered in the development of the final recommendations for the successful resolution of a sick building problem.

ACKNOWLEDGMENTS

The authors acknowledge the analytical assistance and excellent photomicrography skills of William L. Turner Jr. of MVA Scientific Consultants and Mushtaq Khan of Air Quality Sciences Inc.

REFERENCES

American Conference of Governmental Industrial Hygienists (ACGIH). 1999. Developing an investigation strategy. Chapter 2. In: *Bioaerosols: Assessment and Control,* J. M. Macher ed. Cincinnati, OH: American Conference of Governmental Industrial Hygienists,.

American Industrial Hygiene Association (AIHA). 2008. Sampling methods. Chapter 11, 139–151. In: *Recognition, Evaluation and Control of Indoor Mold,* B. Prezant, D. M. Weekes, and J. D. Miller, eds., Fairfax, VA: American Industrial Hygiene Association.

American Society for Testing and Materials International (ASTM). 2003. Standard Practice D6602-03b, Sampling and Testing of Possible Carbon Black Fugitive Emissions or Other Environmental Particulate, or Both. West Conshohocken, PA: American Society for Testing and Materials International.

Brown, R.S., Boltin, W. R., Bandli, B. R., and Millette, J. R. 2007. Light and electron microscopy of mineral wool fibers. *Microscope* 55(1): 37–44.

Edwards, R. W. 1951. Scotch tape-slides for rapid identification of pathogenic fungi. *Laboratory Digest* 15: 8–9.

Farr, D., Bills, G., Chamuris, G., and Rossman A. 1989. *Fungi on plants and plant products in the United States.* St. Paul, MN: American Phytopathological Society Press.

Goldberg, E. D. 1985. *Black Carbon in the Environment: Properties and Distribution.* New York: Wiley & Sons.

Hawksworth, D. L., Kirk, P. M., Sutton, B. C., and Pegler, D. N. 1995. *Ainsworth and Bisby's Dictionary of the Fungi.* Oxon, UK: CABI International.

Institute of Inspection, Cleaning and Restoration Certification (IICRC). 2nd ed. 2008. Standard for Professional Mold Remediation S520, Vancouver, WA: Institute of Inspection, Cleaning and Restoration Certification (IICRC).

Lioy, P. J. 2006. Employing dynamical and chemical processes for contaminant mixtures outdoors to the indoor environment: The implications for total human exposure analysis and prevention. *Journal of Exposure Science and Environmental Epidemiology* 16(3): 1–18.

Lioy, P. J, Freeman, N. C. G., and Millette, J. R. 2002. Institute of Inspection, Cleaning and Restoration Certification. *Environmental Health Perspectives* 110(1): 969–983.

Macher, J. M., ed. 1999. *Bioaerosols: Assessment and Control.* Cincinnati, OH: American Conference of Governmental Industrial Hygienists.

Martyny, J. W., Martinez, K. F., and Morey, P. R. 1999. Source sampling. Chapter 8. In: *Bioaerosols: Assessment and Control*, J. M. Macher ed., Cincinnati, OH: American Conference of Governmental Industrial Hygienists.

McCrone,W. C., and Delly, J. G. 1972. *The Particle Atlas*, Ann Arbor, MI: Ann Arbor Science.

Medalia, A. I., and Rivin, D. 1982. Particulate carbon and other components of soot and carbon black. *Carbon* 20: 481–492.

Millette, J. R. 2001. Early studies characterizing household dirt. *Microscope* 49(4):201–208.

Millette, J. R., and Brown, R. S. 2007. Environmental forensic microscopy. Chapter 13, pp. 611–635. In: *Introduction to Environmental Forensics*, 2nd ed., B. L. Murphy and R. D. Morrison, eds. Elsevier, Amsterdam: Academic Press.

Millette, J. R., Brown, R. S., and Hill, W. B. 2008. Using environmental forensic microscopy in exposure science. *Journal of Exposure Science and Environmental Epidemiology* 18: 20–30.

Millette, J. R., and Few, P. 1998. Indoor carbon soot particles. *Microscope* 46(4): 201–206.

Millette J. R., and Few, P. 2001. Sample collection procedures for microscopical examination of particulate surface contaminants. *Microscope* 49(1): 21–27.

Millette, J. R., Gerber, M., Turner, W., and Hill, W. B. 2007. Investigation of ghosting, a darkening agent on the ceiling. *Microscope* 55(1): 3–7.

Millette, J. R., Hopen, T. J., and Brown, R. S. 1994. Investigating the particulate component in indoor air quality concerns. *Environmental Choices* 3(2):14–16.

Millette J. R., Lioy P. J., Wietfeldt, J., Hopen, T. J., Gipp, M., Padden, T., Singsank, C., and Lepow, W. 2003. A microscopical study of the general composition of household dirt. *Microscope* 51(4): 201–207.

Millette, J. R., Turner, W., Hill, W. B., Few, P., and Kyle, J. P. 2007. Microscopic investigation of outdoor "sooty" surface problems. *Environmental Forensics* 8: 37 –51.

Morey, P. R., Horner, W. E., Epstien, B. L., Worthan, A. G., and Black, M. S. 2000. Indoor air quality in non-industrial occupational environments. Chapter 65, pp. 3149–3241. In: *Patty's Industrial Hygiene*, 5th ed., R. L. Harris (ed.). New York: John Wiley & Sons.

Stevenson, R. 1982. The morphology and crystallography of diesel particulate emissions. *Carbon* 20: 359–365.

Turner, W. L., Millette, J. R., Boltin, W. R., and Hopen, T. J. 2005. A standard approach to the characterization of common indoor dust constituents. *Microscope* 53(4): 169–177.

9 Analytical Practice in Mold Identification and Solutions, Including Measurements and Sampling

Edward Sobek, PhD

CONTENTS

9.1 HISTORICAL PERSPECTIVE ON MOLD SAMPLING

Airborne biological agents have always been a matter of concern in agriculture, biotechnology, industrial settings, and the indoor environment (Deacon, 2006). Each of these environments presents unique exposure based on the nature of the encountered biological agent, the microbial concentrations, the modes of exposure, and the susceptibility of the exposed population. Acceptable levels of airborne microorganisms have not been established, and the sampling methods and analytical techniques employed to assess airborne biological contaminants are varied and nonstandardized (Vesper et al., 2005). Selection of sampling and analytical methods depends upon the nature of the information that is sought; there is no one ideal sampling or analytical method. Combinations of sampling and analytical methods can provide a wide range of data that can be effectively adopted to different environmental settings.

Transmission of airborne microflora occurs both at a local level and over long distances (Griffin, 1972). As a result, numerous studies have been carried out to monitor not only disease and transmission of disease propagules in the aerobiological environment, but also allergens and mycotoxins (Bloom et al., 2009) carrying particles and recent transmission of pollen from genetically modified crops. Actual sampling procedures may involve the passive collection of spores by gravitational deposition or sampling specific volumes of air with "active" spore-trapping devices (Macher, 1999). With most of these techniques, microscopic examination of impaction surfaces is required. If accurate counts are to be obtained, such techniques require considerable amounts of time and expertise. Delays in analysis of samples mean that it is difficult to use results in real time. In addition, the detection of small and nondescript spores is required; species identification is as yet not a realistic option using spore traps.

9.2 CHAIN OF CUSTODY

Laboratories accepting indoor air quality samples require documentation of sample collection, collection protocol, submission, and receipt in the form of a document called a chain of custody or COC (Figure 9.1). This document, in conjunction with a written sample receipt, is retained at the lab to establish an intact, continuous record of physical possession of samples. The COC records an accurate account of the time periods of possession associated with the samples prior to their being received by the lab. It serves to identify all those with primary contact with the samples. All samples require a COC during any type of transport or shipment, and it allows for complete traceability of the samples.

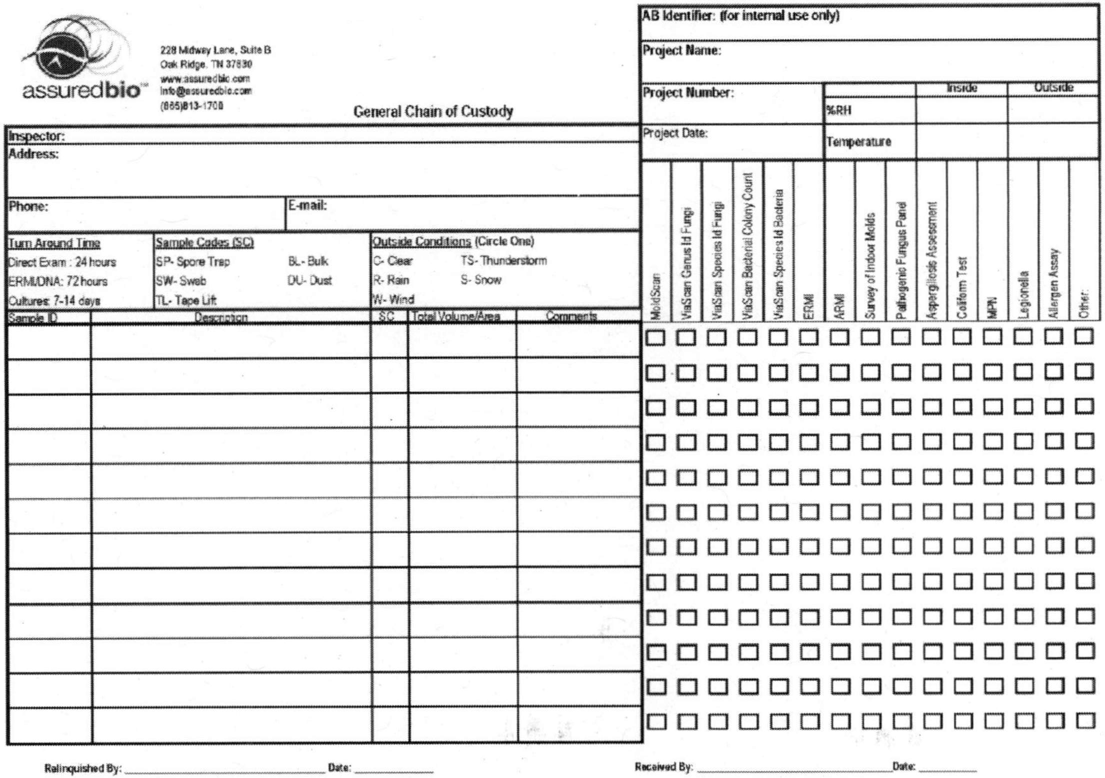

assuredbio™
228 Midway Lane, Suite B
Oak Ridge, TN 37830
www.assuredbio.com
info@assuredbio.com
(865)813-1700

General Chain of Custody

AB Identifier: (for internal use only)
Project Name:
Project Number:
Project Date:

	Inside	Outside
%RH		
Temperature		

Inspector:
Address:
Phone: E-mail:

Turn Around Time	Sample Codes (SC)		Outside Conditions (Circle One)	
Direct Exam : 24 hours	SP- Spore Trap	BL- Bulk	C- Clear	TS- Thunderstorm
ERMI/DNA: 72 hours	SW- Swab	DU- Dust	R- Rain	S- Snow
Cultures: 7-14 days	TL- Tape Lift		W- Wind	

Sample ID	Description	SC	Total Volume/Area	Comments	MoldScan	ViaScan Genus Id Fungi	ViaScan Species Id Fungi	ViaScan Bacterial Colony Count	ViaScan Species Id Bacteria	ERMI	ARMI	Survey of Indoor Molds	Pathogenic Fungus Panel	Aspergillosis Assessment	Coliform Test	MPN	Legionella	Allergen Assay	Other:
					☐	☐	☐	☐	☐	☐	☐	☐	☐	☐	☐	☐	☐	☐	☐
					☐	☐	☐	☐	☐	☐	☐	☐	☐	☐	☐	☐	☐	☐	☐
					☐	☐	☐	☐	☐	☐	☐	☐	☐	☐	☐	☐	☐	☐	☐
					☐	☐	☐	☐	☐	☐	☐	☐	☐	☐	☐	☐	☐	☐	☐
					☐	☐	☐	☐	☐	☐	☐	☐	☐	☐	☐	☐	☐	☐	☐
					☐	☐	☐	☐	☐	☐	☐	☐	☐	☐	☐	☐	☐	☐	☐
					☐	☐	☐	☐	☐	☐	☐	☐	☐	☐	☐	☐	☐	☐	☐
					☐	☐	☐	☐	☐	☐	☐	☐	☐	☐	☐	☐	☐	☐	☐
					☐	☐	☐	☐	☐	☐	☐	☐	☐	☐	☐	☐	☐	☐	☐
					☐	☐	☐	☐	☐	☐	☐	☐	☐	☐	☐	☐	☐	☐	☐
					☐	☐	☐	☐	☐	☐	☐	☐	☐	☐	☐	☐	☐	☐	☐
					☐	☐	☐	☐	☐	☐	☐	☐	☐	☐	☐	☐	☐	☐	☐

Relinquished By: ______ Date: ______ Received By: ______ Date: ______

FIGURE 9.1 General chain of custody: form to establish traceability of samples from the time of sample collection to the time of sample reporting. A COC must accompany every sample and be filled out completely and accurately.

The COC serves other functions as well. It can be utilized to document conditions at the time of sampling; items like temperature, relative humidity, weather conditions, and other pertinent information. Sample descriptions such as the sample number and location of collection are also recorded on the COC. These items can later be used by the property inspector, an industrial hygienist, or the laboratory analyst for reporting and interpreting data. The COC also acts as a service contract between the laboratory and the party submitting the sample. It contains billing, contact information, and turnaround time allowing data to be reported in a timely manner. Without a COC, samples cannot be processed.

9.3 DIRECT EXAM FIELD SAMPLING TECHNIQUES

There are a variety of direct exam methods used to detect indoor molds. They include simple techniques such as swabbing a suspect surface or using clear adhesive tape to lift a sample of mold from a surface. More advanced techniques require calibrated air pumps that are capable of pulling a constant volume of air over time. These pumps are used in conjunction with a spore trap. There are many different makes and models of spore traps, and each comes with a manufacture specification for use.

9.3.1 Spore Sampling Methods

Spore trap analysis is a nonculture-based sampling tool that provides an indication of the kinds and levels of total (viable and nonviable) airborne fungal (mold) spores present in the indoor environment. Direct microscopy is used to analyze the samples, providing both a qualitative and quantitative assessment of spores in the air in a short timeframe. It is used to determine whether the mixture of airborne fungi in a building is normal and typical or indicative of moisture problems.

Air sampling is also an important tool for conducting exposure assessments. Spore trap sampling provides a quicker turnaround time than culture-based analysis when rapid communication of results is essential. Often investigators use nonculturable air sampling as postremediation clearance testing to evaluate effectiveness. This method provides a report on both raw spore count for individual spore types and total concentration expressed as spores per cubic meter (spores/m^3). The proportions of each spore type are also calculated and important groupings of fungal types are summarized to facilitate interpretation of results.

A number of samplers for culturable fungi and for total spore counts are commercially available. These samplers are widely used by allergists, industrial hygienists, and environmental professionals. Sampling for culturable fungi includes agar impact techniques and impingement spore capture (Macher, 1999).

A spore trap air sampling cassette is a sampling device designed for the rapid collection and analysis of a wide range of airborne aerosols. These include fungal spores, pollen, insect parts, skin cell fragments, fibers, and inorganic particulates. Air enters the cassette, the particles become impacted on the sampling substrate, and the air leaves through the exit orifice. The airflow and patented cassette housing is designed in such a way that the particles are distributed and deposited equally on a special glass slide contained in the cassette housing called the trace. This technique has benefits useful for initial site testing, especially if fungal growth is not visible.

Disadvantages of the spore trap method include

- Fungi cannot be fully speciated with this method. For example, *Aspergillus* sp. and *Penicillium* sp. are normally reported together due to the similarities in spore morphology.
- Spore viability cannot be assessed, as it is not possible to differentiate between viable and nonviable.
- Sampling method is cumbersome and noisy.
- Large lab-to-lab variation in identification.
- Methodology not accepted by all within the industry.

9.3.2 Surface Sampling

Direct exam methods for surfaces involve either a tape lift or swab sample. A tape-lift sample is collected by placing clear adhesive tape onto a surface area where mold is suspected to be growing. The sample is placed on a glass slide and sent to the laboratory for identification. A swab sample works much the same way as a tape lift, except that the area suspected to have mold growth is swabbed. The swab is sent to the laboratory for identification. Data reported from tape lifts and swabs for direct exam are qualitative and report the dominant types of mold present on the surface. Usually contamination is given by a genus or group-level identification with a semi-quantitative concentration level such as low, medium, or high.

9.3.3 Bulk Sampling

Bulk sampling for direct exam requires that materials that are suspected to have mold contamination are removed from the building or home and sent to the laboratory for identification. Samples often include cut pieces of drywall, insulation, contaminated framing materials, wood, fabric, or other indoor items. The laboratory will either section the material or use techniques to extract the mold to identify a genus or mold group.

9.4 VIABLE CULTURE FIELD COLLECTION AND SAMPLING TECHNIQUES

Viable culture samples are most often collected using a bioimpact collector or single-stage Andersen sampler. The sampler is precision machined out of an aluminum block that has 400 collecting holes that draw air to capture spores. A petri dish containing an appropriate growth medium is placed inside the sampler. The sampler is connected to an air pump that is calibrated to draw 28.8 L/min of air. A typical sample collection time for most indoor environments is two to three minutes. Clean room and medical areas may need to be sampled longer depending on the type of air filtration used.

9.4.1 Choice of Culture Media

Culture media for viable sampling should consist of a general medium and a selective medium. The general medium will allow all groups of mold to grow, whereas the selective medium will limit some generalist molds while selecting for water intrusion species in the *Aspergillus* and *Penicillium* genera. Malt extract agar is the most common general use medium, while DG18 is commonly used to select water intrusion species. Some Rose Bensal (ROSE) formulations are replacing the MEA and DG18 combination, because they prevent overgrowth without species selectivity. Advances in culture media for mold identification is moving toward a single medium that can allow for identification of generalist molds while also providing resolution to capture the water intrusion species. Some Rose Bengal formulations work well in this respect.

9.4.2 Beware! The Inappropriate Use of Settling Plates

Settling plates are petri dishes that contain a sterilized isolation medium, typically malt extract agar. Settling plates are used in the clean room industry and other industries where there are strict tolerance requirements for microbial contamination. Typically, the settling dish is set out in the area to be sampled and the lid is removed for 15 minutes, replaced, and the plate incubated at an appropriate temperature.

Settling plates are not appropriate for determining indoor mold concentrations in homes and businesses (Macher, 1999). The background mold burden in a typical healthy home will overwhelm a settling plate. Just a few mold spores in a healthy home's dust can give rise to rapidly growing colonies that overwhelm a settling plate in a matter of days. The visual results of rapid colony growth are often interpreted as massive mold contamination. The perception of a fast growing mold such as *Rhizopus stolonifer* can be alarming for an untrained eye, especially a homeowner who may have purchased a settling plate from Home Depot or online. Home test kits that rely on agar-based media in settling plates and incubation by a homeowner should be avoided. They are inaccurate and cause more problems in the long run. Viable sampling is a good method to examine bioaerosol exposure to mold in a home or business. However, this requires a trained inspector, special air pumps, and batch-controlled petri dish media.

9.5 GENUS OR SPECIES IDENTIFICATION AND QUANTIFICATION

The choice of the level of taxonomic resolution in mold identification is determined by the type of sample collected. For example, direct exam analysis via spore trap is relegated to the level of group or genus, whereas as DNA analysis such as mold specific quantitative polymerase chain reaction (MSQPCR) or the Environmental Relative Moldiness Index (ERMI) provide a direct route to species identification (Table 9.1). Both genus and species indicate the degree of genetic relatedness among

TABLE 9.1
Analytical Technique versus Taxonomic Resolution for Section 9.5

Analytical Technique	Initial Taxonomic[a] Resolution[a]	Extended Taxonomic[a] Resolution[a]	Discriminate between *Aspergillus* and *Penicillium* Genera[b]	Standard Laboratory Turnaround Time
MSQPCR/ERMI	Species	Species	Yes	24–72 hours
Spore trap	Group/Genus	Group/Genus	No	24–72 hours
Tape lift	Group/Genus	Group/Genus	Variable	24–72 hours
Swab	Group/Genus	Group/Genus	Variable	24–72 hours
Bulk	Group/Genus	Group/Genus	Variable	24–72 hours
Viable culture	Genus	Species	Yes	7–21 days

[a] Taxonomic resolution is defined as the level of mold classification achievable by an analytic technique. Taxonomy characterization flows from broad groupings to identifying individuals as such: Group > Genus > Species. Both genus and species indicated the degree of genetic relatedness among molds, whereas group is based on morphological similarities only.

[b] *Aspergillus* and *Penicillium* are distinct genera of molds that comprise a significant group of water-intrusion-related mold species. Their spore morphology is very similar (3–5 μm in diameter and round to elliptical in shape), preventing discrimination by some analytical techniques.

molds (Mueller and Schmit, 2007), whereas the group is based on morphological similarities only. Initial taxonomic resolution is the level of reported identification from a first-line analytical technique. Some techniques can be followed up by more discriminating identification such as viable culture. However, some techniques are dead ends because the specimen cannot be removed from the capture matrix, as in the spore trap analysis.

Aspergillus and *Penicillium* are distinct genera of molds that comprise a significant group of water-intrusion-related mold species (Rhodes and Brakhage, 2006; Vesper et al., 2004). Their spore morphology is very similar. Typical spores range in size from 3 to 5 μm in diameter and round to elliptical in morphology. The commonality of size and morphology between *Penicillium* and *Aspergillus* spores prevent discrimination by some analytical techniques such as microscopic examination of spore traps.

The quantity of spores can be determined from most sample matrices (Table 9.2). Spore quantification is relevant to air samples with a defined collection volume and time and also for determining the mold burden of a home from a DNA dust sample. Surface samples such as tape lifts and swabs are limited in their usefulness when spores are quantified. Most often samples from surfaces are collected from mold colonies that have millions of spores and mycelia fragments. In those cases the inspector is looking for positive or negative results as to whether the suspected mold colony is indeed mold and classification of a mold genus. Reporting a million plus spores from a tape lift collected off a mold colony can inflate the perceived size of the contamination issue to the untrained observer, such as a homeowner.

9.6 THE LABORATORY PERSPECTIVE: ENSURING QUALITY ANALYSIS

From an inspector's viewpoint, the choice of a laboratory is extremely important. The lab is not just a place to send samples. The lab is the place to learn how to collect samples, interpret complex reports, purchase supplies, and most importantly discuss projects with professionals that are knowledgeable and trustworthy. Affordable, high-quality analytical techniques and fast turnaround times are the two most important factors for which good labs strive. Great labs do that plus provide superior customer service.

Inspectors should be able to call and discuss reports, optional analyses, and advance techniques with laboratory personnel. A lab that is always trying to sell something rather than help the inspector solve problems is not there to serve the inspector, but rather to boost the bottom line. The lab is a scientific organization first and foremost. Second, it is a business. Yes, labs need to turn a profit but not by sacrificing sound scientific principles. Laboratory directors should have a scientific degree, preferably a PhD in biology, mycology, or microbiology. If they have a PhD ask for their curriculum vitae. Every PhD has one of these and should give it freely. Review their research record. What was their dissertation topic? If the theme was not microbiology related, steer clear. A PhD in nuclear physics and a dissertation on plutonium oscillation principles does little to help one identify *Aspergillus flavus* in a home.

The primary criterion that must be met by any environmental microbiology laboratory that analyzes indoor air samples is accreditation by the American Industrial Hygiene Association (AIHA). The specific program that AIHA offers to mold analysis laboratories is the Environmental Microbiology Laboratory Accreditation Program or EMLAP. AIHA provides a list on its Web site (www.aiha.org) of accredited labs. If a lab is not EMLAP accredited, it is best to avoid it, especially for any project that may go to litigation.

9.6.1 The Laboratory International Standard—ISO 17025:2005

The International Standards Organization (ISO) was formed in 1947 establishing a body of representatives from various countries to set industrial and commercial standards. As of today, it has published over 17,500 standards in industries such as shipping, management, petroleum and gases, medical devices, and laboratory practices. Although there are many other organizations that publish good laboratory practices and offer laboratory accreditation programs, most are based on the standards as outlined by ISO/IEC 17025:2005.

The laboratory standards as covered by ISO require the establishment of a quality assurance program with detailed action steps to ensure quality and repeatable results. Topics outlined by the ISO comprise a wide array of topics including, but not limited to, personnel management, experimental design, equipment maintenance, and customer relations. Each policy is designed to meet customer expectations such as accuracy, reliability, data reporting, data redundancy, and confidentiality. Laboratories that are ISO accredited, or that are accredited by organizations that enforce ISO standards, should have a rigorous quality program in place ensuring good quality work.

9.6.2 Laboratory Accreditations, Proficiency Testing, and Round Robins

There are several different organizations that offer laboratory accreditation programs in environmental microbiology, but none as prestigious as the AIHA. AIHA has established a rigorous program that is in place to ensure that indoor air quality labs are meeting ISO 17025:2005 standards and have a thorough quality control and assurance program. This is managed by several activities designed to evaluate the effectiveness of the quality program in place in the lab.

The lab accreditation process is a long and tedious process. It involves the laboratory quality assurance officer designing a quality program, the publishing of a quality assurance manual, and then putting the program into action. Once this has occurred, AIHA will begin processing the application for accreditation. Laboratory documents and protocols are reviewed. This is done to ensure that one process is in place for each laboratory function. This process is to be tested and validated prior to ever handling a client's samples. A technical advisory panel reviews documentation of methods and method validation, prior to awarding a lab accreditation.

To be accredited, laboratories must show competency. The lab's quality assurance manual and the accreditation policy modules published by AIHA outline qualifications for lab directors, managers, analysts, and technicians. However, does just meeting these standards mean that a scientist is a good one? To show competency, labs are required to undergo proficiency tests to qualify for and maintain a laboratory accreditation.

Repeatability in lab work has been an ongoing theme. AIHA and other accrediting bodies do what is called a round robin to demonstrate repeatability. This exercise involves sample trading between labs. In doing this, each sample is read by two laboratories. It takes the concept of running an in-house replicate sample to the next level. Not only will different analysts analyze the sample, it is analyzed by multiple laboratories at multiple locations. Results from the different analysts should fall within three standard deviations of each other. Failure to demonstrate proficiency and repeatability can lead to a loss of laboratory accreditation.

9.6.3 AIHA EMLAP Program

The AIHA has several programs designed for laboratory accreditation. In the indoor air quality arena, especially when determining a mold or bacterial burden, an accreditation in environmental microbiology is important. This program is called the Environmental Microbiology Laboratory Accreditation Program or EMLAP (www.aihaaccreditedlabs.org/EMLAP). Its policies are written and designed specifically for labs that identify microorganisms commonly found in indoor air or on indoor surfaces. This program serves to protect the integrity of data gathered and reported to mold inspectors, industrial hygienists, and homeowners.

Participation in the EMLAP program is not simply a one-time thing. It is an ongoing process. The laboratory quality program must be enforced daily and the quality assurance manual should be reviewed at least annually. Accredited labs are required to undergo laboratory reaccreditation including a site assessment by an industry professional every two years. Failure to maintain a good quality process can result in loss of laboratory accreditation.

9.6.4 AIHA EMPAT Program

To maintain an AIHA laboratory accreditation, labs must successfully demonstrate proficiency utilizing the methods for which they are accredited. This exercise is done at least three times a year in the form of the Proficiency Analytical Testing (PAT) Program. The AIHA offers several types of PATs including industrial hygiene, lead, beryllium, and asbestos. The most important program for indoor air quality labs is the environmental microbiology PAT or EMPAT (www.aihapat.org/empat).

Laboratories participating in the EMPAT receive blind samples. This means they receive microbial unknowns. Labs that are accredited for culturable fungi or bacteria receive three unknown fungal or bacterial specimens. They are asked to utilize the laboratory's standard operating procedures (just as used on client samples) to culture and identify the unknowns. These can be identified to the genus or species level. The PAT program is time sensitive, so laboratories are required to submit their results within 30 days.

Laboratories that are accredited for direct examination participate in a quarterly EMPAT program designed to simulate scenarios encountered while performing microscopic identifications. A group of 20 Web-based images comprise this test. Labs are asked to identify fungi as specifically as possible and demonstrate the ability to separate items that are nonfungal in nature. Just as the program for culturable fungi and bacteria, this proficiency test is time sensitive. Labs are given minutes to make their identifications. Failure to successfully complete the EMPAT program can lead to a loss of laboratory accreditation.

9.6.5 Laboratory Handling of Samples

The proper handling and data management of samples when out in the field has been discussed. However, what happens in the laboratory? Interestingly, laboratories are required to continue to maintain the chain of custody (COC) as long as the samples remain in house. This means that every step a sample makes through the lab is tracked from the time it is delivered, through all experimental processes, through data reporting, and then sample storage. Responsible labs should be able to locate a sample at any time.

To do this, the policies and procedures held within the quality assurance manual and the standard operating procedures should be followed at all times. The quality assurance manual should specifically address sample management requirements, and the standard operating procedures should outline all methods from sample receipt and log in to data recording. All documents generated while a sample is in house are stored in the sample's file.

Once a sample goes through log in, it is assigned to a laboratory department. This is done based on the type of analysis requested by the inspector or hygienist. DNA samples go to the molecular department, direct exam samples go to microscopy, and culturables go to classical microbiology. The lab file and COC travel with the sample the whole way. The sample is labeled, prepped, and analyzed. The data generated is recorded on a raw data sheet. This may be stored electronically or within the sample file. Finally the data generated is organized and reported to the client. This is done utilizing the data submitted on the COC.

As mentioned earlier, sample handling goes far beyond generating data and formulating a report. Laboratories must manage the samples and the data stored in-house. Accreditation and ISO requirements are such that data redundancy is necessary. This means that laboratories house the original sample (in case reanalysis is requested) and multiple copies of the collected and reported data. Sample project files are kept in hard copies, and reports are stored electronically on and off site for a total of three layers of redundancy. Laboratories normally retain samples, data sheets, and reports for a period of three years. Samples requiring off-site analysis require documentation of such in the form of a completed COC to the receiving lab.

9.6.6 Direct Exam Methods

Direct exam methods involve a simple identification of microorganisms and particulate under the microscope. This can be performed on a sample collected as a spore trap, swab, tape lift, or bulk. Spores that are collected via spore trap can be physically identified, counted, and quantified. All other sampling methods report solely by identification and are reported in terms of low, medium, and high. Direct examination is a great technique to use when looking for a rapid identification of the mold genera present. Some microscopic methods can also be used for basic bacteria classifications, however, testing for bacteria using a spore trap is not recommended. They are simply too small to be effectively captured in the collection media.

Direct examination of a spore trap reports (Table 9.2) several pieces of information about each individual sample. First is the raw count. This number represents the organisms identified by the analyst and the physical number of each organism collected in the sample. The quantification of spores per cubic meter accompanies this number. This number is a calculation that is based on the total volume of air collected in the sample. Normally this is calculated for each individual spore classification and for the total spore count. Other information that is normally included on a direct exam report is the limit of detection and the background particulate density. The limit of detection is the lowest number of spores that can be detected based on the volume of the sample collected. The background particulate density is a rating indicating the presence of airborne particulates other than mold (pollen, dander, insects, etc.). As the background particulate density increases, the ability to visually detect smaller mold spores (such as *Penicillium*/*Aspergillus* or *Acremonium*) decreases. This information can be valuable in determining the reliability of the spore counts. Always be mindful when utilizing air sampling techniques, as there are currently no national guidelines or standards for the acceptable number of mold spores present in indoor air.

It has become the industry standard for professionals to do a comparison between samples collected on the interior of a home or building and an outside control sample. The fungal spores recovered indoors should be similar in type and number to

TABLE 9.2
Example of Quantification of Spores from Spore Trap and Swab from a Direct Examination Report

Assured Biotechnologies Corporation
Direct Examination Analysis

228 Midway Lane, Suite B
Oak Ridge, TN 37830

www.assuredblo.com

Inspector:	Certified Mold Inspector	Date Collected:	09/25/09
Project	A Moldy House	Date Received:	09/28/09
Job Number:	123456	Date Reported:	09/28/09
Assured Bio Identifier:	CMI092809-3	Analyst:	A Lab Analyst

Selected References

1. Alexopoulos, C. J. and C.W. Mims. 1979. *Introductory Mycology, Third Edition.* John Wiley & Sons, New York, New York.
2. Barron, G. L. 1968. *The Genera of Hyphomycetes from Soil.* Robert E. Krieger Pub. Co., Malabar, Florida.
3. Ellis, M.B. 1971. *Dematiaceous Hyphomycetes.* CAB International, Wallingford Oxon OX10 8DE, UK.
4. Ellis, M.B. 1976. *More Dematiaceous Hyphomycetes.* CAB International, Wallingford Oxon OX10 8DE, UK.
5. Hanlin, R.T. 1990. *Illustrated Genera of Ascomycetes.* APS Press, St. Paul, Minnesota.
6. Hanlin, R.T. 1998. *Illustrated Genera of Ascomycetes.* II. APS Press, St. Paul, Minnesota.
7. Hanlin R.T. and M. Ulloa. 1988. *Atlas of Introductory Mycology, Second Edition.* Hunter Textbooks, Inc., Winston-Salem, North Carolina.
8. Kiffer E. and M. Morelet. 2000. *The Deuteromycetes: Mitosporic Fungi: Classification and Generic Keys.* Science Publishers, Inc., Enfield, New Hampshire.
9. Macher, J., Ed. 1999. *Bioaerosols Assessment and Control.* ACGIH, Cincinnati, Ohio.
10. Morris, E.F. 1963. *The Synnematous Genera of the Fungi Imperfecti.* Western Illinois University Publication, Macomb, Illinois.
11. Nelson, P.E., Toussoun, T.A. and W.F.O. Marasas. 1983. *Fusarium Species: An Illustrated Manual for Identification.* Pennsylvania State University Press, University Park and London.
12. Samson R.A., Hoekstra E.S., Frisvad J.C. and O. Filtenborg, Ed. 2002. *Introduction to Food and Airborne Fungi.* Ponsen and Looyen, Wageningen, The Netherlands.
13. Wistreich G.A. 1997. *Microbiology Laboratory: Fundamentals and Applications.* Prentice Hall, Upper Saddle River, New Jersey.

1 of 10

Limitations

ASSURED BIOTECHNOLOGIES CORP. MAKES NO WARRANTIES AND EXPRESSLY DISCLAIMS THE IMPLIED WARRANTIES OF MERCHANTABILITY AND FITNESS FOR PURPOSE. INSPECTOR ACKNOWLEDGES THAT ASSURED BIOTECHNOLOGIES CORP. HAS NOT INSPECTED THE SUBJECT PROPERTY AND THAT INSPECTOR IS SOLEY RESPONSIBLE FOR CHOOSING THE SITES FOR PLACEMENT OF THE SPORE TRAPS. ASSURED BIOTECHNOLOGIES CORP. SHALL NOT BE LIABLE TO INSPECTOR FOR ANY INCIDENTAL, CONSEQUENTIAL, SPECIAL OR PUNITIVE DAMAGES OF ANY KIND OR NATURE, INCLUDING, WITHOUT LIMITATION, ANY DAMAGES TO PROPERTY OR PERSONAL INJURY CAUSED BY WATER INTRUSION, MOLD OR MOISTURE IN THE PREMISES, WHETHER SUCH LIABILITY IS ASSERTED ON THE BASIS OF CONTRACT, TORT, OR OTHERWISE, EVEN IF ASSURED BIOTECHNOLOGIES CORP. HAS BEEN WARNED OF THE POSSIBILITY OF SUCH LOSS OR DAMAGE. UNDER NO CIRCUMSTANCES SHALL ASSURED BIO BE LIABLE FOR DAMAGES UNDER OR ARISING OUT OF THIS REPORT IN AN AMOUNT EXCEEDING THE AMOUNT PAID BY INSPECTOR TO ASSURED BIOTECHNOLOGIES CORP. FOR THIS ANALYSIS AND REPORT. THIS REPORT IS FOR THE SOLE USE OF INSPECTOR AND CREATES NO THIRD PARTY BENEFICIARIES OR RIGHTS HEREUNDER.

Methods of Analysis

Assured Biotechnologies Corp. uses the following Standard Operating Procedures for the analysis of samples:

Spore Traps	- Assured Biotechnologies Corp. Document Number 105
Swabs	- Assured Biotechnologies Corp. Document Number 106
Tape Lifts	- Assured Biotechnologies Corp. Document Number 107
Bulk Material	- Assured Biotechnologies Corp. Document Number 108

Interpretation of Spore Trap Results

The Interior vs. the Outside Control: There are currently no national guidelines or standards for the acceptable number of mold spores present in indoor air. It has become the industry standard for professionals to do a comparison between samples collected on the interior of a home or building and the outside control. The fungal spores recovered indoors should be similar in type and number to those collected outside. When this occurs the area is considered to be in equilibrium or is seen as a normal indoor environment. It is very common for multiple samples to be collected on the interior. This is done to pinpoint any problem areas where air quality may be abnormal.

Limit of Detection: This number is the lowest number of spores that can be detected based on the volume of the sample collected.

Background Particulate Density: This rating indicates the presence of airborne particulates other than mold (pollen, dander, insects, ect.). As the Background Particulate Density increases, the ability to visually detect smaller mold spores (such as *Penicillium/Aspergillus* or *Acremonium*) decreases. The Level of Debris can be interpreted using the following scale:

Low	-Very little particulate present. Virtually no spores undetectable.
Low-Medium	-Little particulate present. 97%likelihood that all spores have been counted.

Medium	-Moderate level of particulate present. 95% likelihood that all spores have been counted.
Medium-High	-Increased level of particulate present. 75% likelihood that all spores have been counted.
High	-Very heavy particulate. Less than a 75% likelihood that all spores have been counted.

Raw Count: This number is the total number of fungal spores counted under the microscope.

Total Mold Spores: This is an estimate of the fungal spores present per cubic meter of air sampled within that particular sampling location. This number is derived by multiplying the total spores counted on the spore trap by a conversion factor involving the volume of air sampled.

Interpretation of Tape Lift, Swab, or Bulk Material Results

Assured Bio. quantifies the presence of mold spores detected on a tape lift, swab, or bulk sample using the following scale:

Low	-Less than 1/3 of the microscopic field of view is obscured by the identified mold spore.
Moderate	-Between 1/3 and 2/3 of the microscopic field of view is obscured by the identified mold spore.
High	-Over 2/3 of the microscopic field of view is obscured by the identified mold spore.

ABC Identification Number:	CMI092809-3-1	CMI092809-3-4
Sample Identification Number:	1661457	1669978
Date Collected:	09/25/09	09/25/09
Description:	Kitchen	Outside
Sample Type:	Spore Trap	Spore Trap
Sample Condition:	Intact	Intact
Comments:		
Volume/Area Sampled:	25	25
Analytical Sensitivity:	40	40

Spore Identifications	**Raw Count**	**Spores/m³**	**Raw Count**	**Spores/m³**
Acremonium-like	ND	BDL	ND	BDL
Alternaria	ND	BDL	ND	BDL
Arthrinium	ND	BDL	ND	BDL
Aureobasidium	ND	BDL	ND	BDL
Botrytis	ND	BDL	ND	BDL
Cercospora	ND	BDL	ND	BDL
Chaetomium	ND	BDL	ND	BDL
Cladosporium	4	160	1	40
Coprinus	ND	BDL	1	40
Curvularia	ND	BDL	1	40
Drechslera/Bipolaris	ND	BDL	ND	BDL
Epicoccum	ND	BDL	ND	BDL
Fusarium	ND	BDL	ND	BDL
Ganoderma	ND	BDL	11	440
Memnoniella	ND	BDL	ND	BDL
Nigrospora	ND	BDL	ND	BDL
Penicillium/Aspergillus-like	946	37,840	ND	BDL
Pithomyces	ND	BDL	ND	BDL
Scopulariopsis	ND	BDL	ND	BDL
Spegazzinia	ND	BDL	ND	BDL
Stachybotrys	ND	BDL	ND	BDL
Tetraploa	ND	BDL	ND	BDL
Torula	ND	BDL	ND	BDL
Trichoderma	ND	BDL	ND	BDL
Ulocladium	ND	BDL	ND	BDL
Wallemia	ND	BDL	ND	BDL
Ascomycetes-unspecified	ND	BDL	15	600
Basidiomycetes-unspecified	ND	BDL	1	40
Hyphomycetes-unspecified	ND	BDL	ND	BDL
Rusts/Smuts/Myxomycetes	ND	BDL	ND	BDL
Zygomycetes-unspecified	ND	BDL	ND	BDL
Total spore count	950	38,000	30	1,200
Hyphae	ND	BDL	ND	BDL
Background particulate density	Low		Medium	

ABC Identification Number:	CMI092809-3-2	CMI092809-3-4
Sample Identification Number:	1669997	1669978
Date Collected:	09/25/09	09/25/09
Description:	Guest Bedroom(s)	Outside
Sample Type:	Spore Trap	Spore Trap
Sample Condition:	Intact	Intact
Comments:		
Volume/Area Sampled:	25	25
Analytical Sensitivity:	40	40

Spore Identifications	**Raw Count**	**Spores/m³**	**Raw Count**	**Spores/m³**
Acremonium-like	ND	BDL	ND	BDL
Alternaria	ND	BDL	ND	BDL
Arthrinium	ND	BDL	ND	BDL
Aureobasidium	ND	BDL	ND	BDL
Botrytis	ND	BDL	ND	BDL
Cercospora	ND	BDL	ND	BDL
Chaetomium	1	40	ND	BDL
Cladosporium	1	40	1	40
Coprinus	ND	BDL	1	40
Curvularia	ND	BDL	1	40
Drechslera/Bipolaris	1	40	ND	BDL
Epicoccum	ND	BDL	ND	BDL
Fusarium	ND	BDL	ND	BDL
Ganoderma	ND	BDL	11	440
Memnoniella	ND	BDL	ND	BDL
Nigrospora	ND	BDL	ND	BDL
Penicillium/Aspergillus-like	1,166	46,640	ND	BDL
Pithomyces	1	40	ND	BDL
Scopulariopsis	ND	BDL	ND	BDL
Spegazzinia	ND	BDL	ND	BDL
Stachybotrys	ND	BDL	ND	BDL
Tetraploa	ND	BDL	ND	BDL
Torula	ND	BDL	ND	BDL
Trichoderma	ND	BDL	ND	BDL
Ulocladium	ND	BDL	ND	BDL
Wallemia	ND	BDL	ND	BDL
Ascomycetes-unspecified	ND	BDL	15	600
Basidiomycetes-unspecified	ND	BDL	1	40
Hyphomycetes-unspecified	ND	BDL	ND	BDL
Rusts/Smuts/Myxomycetes	ND	BDL	ND	BDL
Zygomycetes-unspecified	ND	BDL	ND	BDL
Total spore count	1,170	46,800	30	1,200
Hyphae	ND	BDL	ND	BDL
Background particulate density	Low-medium		Medium	

5 of 10

ABC Identification Number:	CMI092809-3-3
Sample Identification Number:	S925
Date Collected:	09/25/09
Description:	Kitchen Wall
Sample Type:	Swab
Sample Condition:	Intact
Comments:	

Spore Identifications	**Spore Concentration**
Acremonium	ND
Acremonium-like	ND
Alternaria	ND
Arthrinium	ND
Aspergillus	ND
Aureobasidium	ND
Botrytis	ND
Cercospora	ND
Chaetomium	Medium
Cladosporium	ND
Coprinus	ND
Curvularia	ND
Drechslera/Bipolaris	ND
Epicoccum	ND
Fusarium	ND
Ganoderma	ND
Memnoniella	ND
Nigrospora	ND
Paecilomyces	ND
Penicillium	ND
Penicillium/Aspergillus-like	Medium
Pithomyces	ND
Scopulariopsis	ND
Spegazzinia	ND
Stachybotrys	High
Tetraploa	ND
Torula	ND
Trichoderma	ND
Ulocladium	ND
Wallemia	ND
Ascomycetes-unspecified	Low
Basidiomycetes-unspecified	ND
Hyphomycetes-unspecified	ND
Rusts/Smuts/Myxomycetes	ND
Zygomycetes-unspecified	ND
Miscellaneous structures	ND
Hyphae	Present
Clamydospores	ND
Perithecia	ND
Sclerotia	ND
Background particulate density	Low-Medium

6 of 10

AssuredBio™

Frequently Occurring Fungi

Acremonium	This organism grows on dead plant material and soil. For growth indoors, it requires very wet conditions. The pathology to humans on exposure includes allergies (eg. hay fever, asthma), pneumonia, and subcutaneous infection.
Alternaria	This can be found on dead and dying plant material. It is easily blown by wind and found in house dust, carpets, textiles, and horizontal surfaces indoors. It can be considered a water impact mold. The pathology to humans on exposure includes allergies and asthma. Other diseases linked to *Alternaria* include mycotic keratitis, skin infections, and osteomyelitis.
Ascomycetes	This group includes over 3,000 species of fungi which mature in a sack-like structure. They are found everywhere in nature. This group includes *Chaetomium* and *Ascotricha* which can frequently found growing indoors on damp substrates. The pathology to humans on exposure is mostly allergenic.
Aspergillus	This can be found growing on forage products, grains, nuts, organic debris and water damaged organic building materials. Pathology to humans includes asthma, but it is less allergenic than other molds. Infections from Aspergillus happen mostly to persons with compromised immune systems. Aspergillosis is the second most common fungal infection requiring hospitalization in US.
Aspergillus/Penicillium	This group of fungal spores includes both the *Aspergillus* and *Penicillium* genera. This is because microscopically the two can not be differentiated unless conidiophores (fungal fruiting bodies) are present in the sample. These organisms are very common in the environment, however, an elevated presence can be indicative of a water intrusion event.
Basidiomycetes	This group of fungal spores originates from mushrooms and plant pathogens. They are found in gardens, forests, and woodlands, but basidiomycetes can grow indoors. *Serpula lacrimans* or "dry rot" and other fungi cause white and brown wood rot. They grow and destroy the structural wood of buildings. The pathology to humans on exposure is mostly allergenic (eg. hay fever, asthma).

Frequently Occurring Fungi (Cont.)

Chaetomium	This organism grows on substrates containing cellulose, including paper and plant compost. It is found commonly on damp sheetrock paper. Spores are distinctively shaped and resemble a lemon or a football and mature in a sack-like structure called perithecia. The pathology to humans on exposure includes Type I and III allergens and can produce a mycotoxin shown to cause kidney and liver damage in laboratory animals.
Cladosporium	This genus grows on living and dead plant material, soil, paint, and textiles. It can be found growing in dirty refrigerators and on air conditioning vents. It grows especially well in reservoirs where condensation collects. Often it is found on the surface of fiberglass duct liners in the interior of supply ducts. The spores are generally dispersed by the wind. Water conditions of *Cladosporium* include houses in damp areas with poor ventilation. It can also be found living on textiles or paper under humid conditions and on moist window frames. Human exposure is rarely pathogenic, but can cause skin lesions, sinusitis, and pulmonary infections. Airborne spores can be significant allergens.
Curvularia	This grows on plant debris and soil. It is a facultative plant pathogen of tropical or subtropical plants and can grow indoors on a variety of substrates, usually under high humidity. The pathology to humans on exposure includes allergies (eg. hay fever, asthma). It is a relatively common cause of allergic fungal sinusitis
Hyphal Fragments	This is the growing part of fungi. Hyphal fragments present in air samples can be indicative of actively occurring mold growth within the indoor environment.
Memnoniella	This organism is closely related to *Stachybotrys* and grows on soil, many types of plants, and trees. It is associated often with water intrusion and can grow indoors on many different substrates. It is found frequently in conjunction with *Stachybotrys.*
Myxomycetes/Rusts/Smuts	These types of fungi are typically found outdoor. Rusts and smuts are often considered plant pathogens or parasites, while myxomycetes are slime molds. These spores are difficult to differentiate microscopically and normally are quantified together.

Frequently Occurring Fungi (Cont.)

Penicillium	This fungus grows on materials such as soil, food cellulose, paint, grains, and compost piles. Spores are commonly found in carpet, wallpaper, and in interior fiberglass duct insulation. *Penicillium* can grow indoors in water-damaged buildings on wallpaper, wallpaper glue, decaying fabrics, moist chipboards, and behind paint. The pathology to humans includes allergies (eg. hay fever, asthma), moldy wall hypersensitivity, and hypersensitivity pneumonitis.
Stachybotrys	This can be found growing on sheet rock, paper, ceiling tiles, cellulose containing insulation backing, and wallpaper. It is a sooty black fungus occasionally accompanied by a thick, mass of white mycelia Conditions for growth include areas subject to temperature fluctuations that also have a relative humidity above 55%. The pathology to human exposure may include allergies, dermatitis, cough, rhinitis, nose bleeds, cold and flu symptoms, headache, general malaise and fever, and diarrhea. It produces mycotoxins which are extremely potent. Toxins produced by the fungus may suppress the immune system-affecting the lymphoid tissue and the bone marrow. Exposure via inhalation, ingestion, or dermal/skin should be avoided.
Ulocladium	This organism grows on plant materials and soils, rotten woods, paper, textiles, and water-damaged building materials. It can be found in dust and air samples. Growth indoors is widespread. It has a high water requirement. The pathology to humans on exposure includes allergies (eg. hay fever, asthma). When this organism is in the presence of *Alternaria*, symptoms may compound.

assuredbio™
228 Midway Lane, Suite B
Oak Ridge, TN 37830
www.assuredbio.com
info@assuredbio.com
(865)813-1700

General Chain of Custody

AB Identifier: (for internal use only) CMI092809-3

Project Name: A Moldy House

Project Number: 123456

Project Date: 9/27/09

	Inside	Outside
%RH	55	60
Temperature	72	89

Inspector: Certified Mold Inspector
Address: 123 Main Street
Any Town, NJ

Phone: (123) 456-7890 **E-mail:** moldreport@moldyhouse.com

Turn Around Time	Sample Codes (SC)		Outside Conditions (Circle One)	
Direct Exam : 24 hours	SP- Spore Trap	BL- Bulk	C- Clear	TS- Thunderstorm
ERMI/DNA: 72 hours	SW- Swab	DU- Dust	R- Rain	S- Snow
Cultures: 7-14 days	TL- Tape Lift		W- Wind	

Sample ID	Description	SC	Total Volume/Area	Comments	MoldScan	ViaScan Genus Id Fungi	ViaScan Species Id Fungi	ViaScan Bacterial Colony Count	ViaScan Species Id Bacteria	ERMI	ARMI	Survey of Indoor Molds	Pathogenic Fungus Panel	Aspergillosis Assessment	Coliform Test	MPN	Legionella	Allergen Assay	Other
1661457	Kitchen	SP	25L		■	☐	☐	☐	☐	☐	☐	☐	☐	☐	☐	☐	☐	☐	☐
1669997	Guest Bedroom (s)	SP	25L		■	☐	☐	☐	☐	☐	☐	☐	☐	☐	☐	☐	☐	☐	☐
S925	Kitchen Wall	SW	N/A		■	☐	☐	☐	☐	☐	☐	☐	☐	☐	☐	☐	☐	☐	☐
1669978	Outside	SP	25L		■	☐	☐	☐	☐	☐	☐	☐	☐	☐	☐	☐	☐	☐	☐
					☐	☐	☐	☐	☐	☐	☐	☐	☐	☐	☐	☐	☐	☐	☐
					☐	☐	☐	☐	☐	☐	☐	☐	☐	☐	☐	☐	☐	☐	☐
					☐	☐	☐	☐	☐	☐	☐	☐	☐	☐	☐	☐	☐	☐	☐
					☐	☐	☐	☐	☐	☐	☐	☐	☐	☐	☐	☐	☐	☐	☐
					☐	☐	☐	☐	☐	☐	☐	☐	☐	☐	☐	☐	☐	☐	☐
					☐	☐	☐	☐	☐	☐	☐	☐	☐	☐	☐	☐	☐	☐	☐
					☐	☐	☐	☐	☐	☐	☐	☐	☐	☐	☐	☐	☐	☐	☐
					☐	☐	☐	☐	☐	☐	☐	☐	☐	☐	☐	☐	☐	☐	☐

Relinquished By: Certified Mold Inspector Date: 9/27/09

Received By: Laboratory Analyst Date: 9/28/09

those collected outside. When this occurs the area is considered to be in equilibrium or is seen as a normal indoor environment. It is very common for multiple samples to be collected on the interior. This is done to pinpoint any problem areas where air quality may be abnormal.

Direct examination from a swab, a tape lift, or a bulk sample is an excellent method for determining the presence of visible mold. Although the data reported is not numeric, this technique does a great job in differentiating the various genera of molds and is extremely reliable in determining the presence or absence of mold. This method has also proven helpful in linking the origin of contamination to the results found utilizing DNA and traditional methods. Results from these types of surface samples are reported in concentrations of high, medium, or low.

High concentrations reflect that over two-thirds of the microscopic field of view is obscured by the identified mold spore. Medium reflects between one-third and two-thirds of the microscopic field of view is obscured by the identified mold spore, and low concentrations reflect less than one-third of the microscopic field of view is obscured by the identified mold spore.

Interpretations and comments from the analyst are documented on the raw data sheet for the sample. These are retained and utilized to build a laboratory report. Lab reports should always be presented in a PDF format, ensuring that the data is original and unaltered. They should also be reviewed for errors and returned to the client with a copy of the COC.

9.6.7 Viable Culture Methods

Viable culture methods identify living microorganisms such as bacteria and fungi that are found during indoor air quality studies. Viable methods can be performed on samples collected as a swab, bulk, air, or water sample (Macher, 1999; Zorman and Jersek, 2008). The microbes collected in the sample are grown on selective media which expedite their growth. Just as individual people are picky, bacteria and fungi are too. They like to consume different foods and like to grow under different environmental conditions. Knowing this, a viable culture method can be designed to identify specific organisms based on the organism's individual preferences. For example, when looking for fungal contaminants, a malt-based media (MEA) and a temperature around 27°C are ideal. Bacteria like to grow on trypic soy agar (TSA) at 37°C, so the two different organisms can be easily targeted. This practice can be carried down to even a species level.

Once viable samples are received at the lab, they are separated for identification based on the contaminant in question. Regardless of the target in question, they are prepared the same. Bulk samples and swab samples are diluted. This is normally done in increments of a log. Doing this allows the analyst to prepare each dilution as a culture hopefully yielding a plate that is easy to physically count. Air samples do not undergo this process; their collection method is designed to yield manageable plates.

The raw counts obtained from an agar plate are called colony counts. They are reported in what is called CFUs or colony forming units. This is because one individual bacterial spore or mold spore will generate the growth of a colony making them easy to quantify. It is recommended in *Bioaerosols: Assessment and Control*

(Macher, 1999) that when performing plate counts "25 to 250 bacterial colonies and 10 to 60 fungal colonies are considered optimal for accurate counting and identification of CFUs on standard 100-mm plates."

Lab reports (Table 9.3) from viable samples include raw colony counts in CFUs and the quantifications of CFUs per unit of measure. This can be done for total fungi and bacteria, genus or species specific. Swabs are reported based on the surface area sampled (e.g., one square inch). Bulk samples are reported by the weight of the sample in grams. Air samples are reported in cubic meters. Water samples are reported in milliliters. Other information included on viable reports are the minimum reporting limit and the analytical sensitivity. The AIHA defines the minimum reporting limit in AIHA LQAP Policy Document–Module 9 as "the minimum concentration of an analyte that, in a given matrix and with a specific method, has a 99 percent probability of being identified, qualitatively or quantitatively measured, and reported to be greater than zero." Analytical sensitivity is defined by the AIHA LQAP Policy Document–Module 9 as "the lowest concentration that can be detected by the method, based upon the amount or portion of sample analyzed." Both of these calculations are significant because they define the boundaries of the test. The minimum reporting limit is the fewest colonies that can be counted on a 1:1 dilution agar plate. The analytical sensitivity is the fewest number of CFUs that can be extrapolated from a collected specimen.

9.8 DNA DETECTION OF ENVIRONMENTAL MICROBES: MOVING TOWARD THE INDUSTRY'S FIRST STANDARD

The rapid advancement and precipitous drop in cost of DNA-based analytical tools has changed the landscape of indoor mold detection forever. DNA analysis, once limited to the pharmaceutical industry or in National Institutes of Health (NIH)-funded research laboratories, has moved into mainstream commercial laboratories. Unlike most other mold analyses, DNA mold detection is very similar to analytical chemical techniques such as gas chromatography-mass spectroscopy (GC-MS) or high performance liquid chromatography (HPLC); particularly, in terms of incorporating reference standards, controls, and blanks (Haugland et al., 2004).

The problem with other mold laboratory methods is the inherent variation of mold spores and mold cultures. A species of mold will grow and look different if grown on different culture media or the same medium at different temperatures. No two mold spores collected from the same mold colony look alike. Spores may be similar, but the differences are sometimes enough to cause confusion in identification. DNA circumscribes the inherent variation of mold morphology by identifying molds based on their genetic code or DNA sequence. DNA probes, like chemical reference materials, are extremely specific and characterized. Moreover, DNA analysis takes advantage of the biological specificity required when organisms undergo cell replication. If major errors occur in DNA replication when a cell divides, an organism mutates and dies. Hence, the error rate is extremely low and DNA replication is considered a high-fidelity reaction that ensures errors do not occur (Heitman et al., 2006). DNA mold detection has all the attributes of a laboratory method on which standards can be built. Most important, DNA analysis can be consistently repeated within and among laboratories with nonsignificant variation in data output.

TABLE 9.3
Viable Culture Report

assuredbio™

Assured Biotechnology Corporation
ViaScan Analysis

228 Midway Lane, Suite B — Phone: (865) 813-1700
Oak Ridge, TN 37830 — Fax: (865) 813-1705

www.assuredblo.com

Inspector:	Certified Mold Inspector	Date Collected:	6/7/2009
Project	A Moldy House	Date Received:	6/8/2009
Job Number:	123456	Date Reported:	6/18/2009
Assured Bio Identifier:	CMI060809-2	Analyst:	E Sobek Ph. D.

Selected References

1. Alexopoulos, C.J. and C.W. Mims. 1979. *Introductory Mycology, Third Edition.* John Wiley & Sons, New York, New York.
2. Barron, G.L. 1968. *The Genera of Hyphomycetes from Soil.* Robert E. Krieger Pub. Co., Malabar, Florida.
3. Ellis, M.B. 1971. *Dematiaceous Hyphomycetes.* CAB International, Wallingford Oxon OX10 8DE, UK.
4. Ellis, M.B. 1976. *More Dematiaceous Hyphomycetes.* CAB International, Wallingford Oxon OX10 8DE, UK.
5. Hanlin, R.T. 1990. *Illustrated Genera of Ascomycetes.* APS Press, St. Paul, Minnesota.
6. Hanlin, R.T. 1998. *Illustrated Genera of Ascomycetes.* II. APS Press, St. Paul, Minnesota.
7. Hanlin R.T. and M. Ulloa. 1988. *Atlas of Introductory Mycology, Second Edition.* Hunter Textbooks, Inc., Winston-Salem, North Carolina.
8. Kiffer E. and M. Morelet. 2000. *The Deuteromycetes: Mitosporic Fungi: Classification and Generic Keys.* Science Publishers, Inc., Enfield, New Hampshire.
9. Macher, J., Ed. 1999. *Bioaerosols: Assessment and Control.* ACGIH, Cincinnati, Ohio.
10. Morris, E.F. 1963. *The Synnematous Genera of the Fungi Imperfecti.* Western Illinois University Publication, Macomb. Illinois.
11. Nelson, P.E., Toussonn, T.A. and W.F.O Marasas. 1983. *Fusarium Species: An Illustrated Manual for Identification.* Pennsylvania State University Press, University Park and London.
12. Samson R.A., Hoekstra E.S., Frisvad J.C. and O. Filtenborg, Ed. 2002. *Introduction to Food and Airborne Fungi.* Ponsen and Looyen, Wageningen, The Netherlands.
13. Wistreich G.A. 1997. *Microbiology Laboratory: Fundamentals and Applications.* Prentice Hall, Upper Saddle River, New Jersey.

Limitations

ASSURED BIOTECHNOLOGY CORP. MAKES NO WARRANTIES AND EXPRESSLY DISCLAIMS THE IMPLIED WARRANTIES OF MERCHANTABILITY AND FITNESS FOR PURPOSE. INSPECTOR ACKNOWLEDGES THAT ASSURED BIOTECHNOLOGY CORP. HAS NOT INSPECTED THE SUBJECT PROPERTY AND THAT THE INSPECTOR IS SOLEY RESPONSIBLE

1 of 4

FOR CHOOSING THE LOACTION OF SAMPLE COLLECTION. ASSURED BIOTECHNOLOGY CORP. SHALL NOT BE LIABLE TO INSPECTOR FOR ANY INCIDENTAL, CONSEQUENTIAL, SPECIAL OR PUNITIVE DAMAGES OF ANY KIND OR NATURE, INCLUDING, WITHOUT LIMITATION, ANY DAMAGES TO PROPERTY OR PERSONAL INJURY CAUSED BY WATER INTRUSION, MOLD OR MOISTURE IN THE PREMISES, WHETHER SUCH LIABILITY IS ASSERTED ON THE BASIS OF CONTRACT, TORT, OR OTHERWISE, EVEN IF ASSURED BIOTECHNOLOGY CORP. HAS BEEN WARNED OF THE POSSIBILITY OF SUCH LOSS OR DAMAGE. UNDER NO CIRCUMSTANCES SHALL ASSURED BIOTECHNOLOGY CORP. BE LIABLE FOR DAMAGES UNDER OR ARISING OUT OF THIS REPORT IN AN AMOUNT EXCEEDING THE AMOUNT PAID BY THE INSPECTOR TO ASSURED BIOTECHNOLOGY CORP. FOR THIS ANALYSIS AND REPORT. THIS REPORT IS FOR THE SOLE USE OF THE INSPECTOR AND CREATES NO THIRD PARTY BENEFICIARIES OR RIGHTS HEREUNDER.

Methods of Analysis

Assured Biotechnology Corporation uses the following Standard Operating Procedures for the analysis of samples:

ViaScan/ Culturable Bacteria from Bulk Material: 125
ViaScan/ Culturable Bacteria from a Swab: 126
ViaScan/ Culturable Bacteria from an Air Sample: 138
Bacterial Species ID for Dominant Organisms: 117, 118, 119, 120
Bacteria Species Id of Enteric Gram Negative Bacteria: 142

ViaScan/ Culturable Fungi from Bulk Material: 122
ViaScan/ Culturable Fungi from a Swab: 124
ViaScan/ Culturable Fungi from an Air Sample: 138
Fungal Species ID for Dominant Organisms: 117, 118, 119, 120

Reporting Limits

Minimum Reporting Limit: The American Industrial Hygiene Association defines this term in AIHA LQAP Policy Document – Module 9 as "The minimum concentration of an analyte that, in a given matrix and with a specific method, has a 99 percent probability of being identified, qualitatively or quantitatively measured, and reported to be greater than zero."

Analytical Sensitivity: The American Industrial Hygiene Association defines this term in AIHA LQAP Policy Document – Module 9 as "The lowest concentration that can be detected by the method, based upon the amount or portion of sample analyzed."

Additional Comments

The analytical data included in this report reflect only the conditions of the material sampled and submitted to the laboratory for analysis at the time of collection. The results included in this report may not be used for past or future environmental conditions.

Assured Biotechnology Corporation utilizes the standard outlined in *Bioaerosols: Assessment and Control* by J. Macher when making reliable interpretations. It states, "In general, 25 to 250 bacterial colonies and 10 to 60 fungal colonies are considered optimal for accurate counting and identification of CFU's on standard 100-mm plates."

assuredbio™

Sample Number:	CMI060809-2-1	Incubation Temperature:	27 C
Sample ID:	Basement Air Sample 1	Sample Volume:	55.19 L
Sample Condition:	Intact	Sample Type:	
Minimum Reporting Limit:	1 CFU	Analytical Sensitivity:	18 CFU/cubic meter

Colony Identifications	Colony Forming Units Counted	Colony Forming Units/cubic meter
Fusarium solani	2	36
Cladosporium sphaerospermum	1	18
Other Fungi	3	54

The total colony forming units per cubic meter is 108.

Sample Number:	CMI060809-2-2	Incubation Temperature:	27 C
Sample ID:	Air Sample Basement 2	Sample Volume:	55.19 L
Sample Condition:	Intact	Sample Type:	
Minimum Reporting Limit:	1 CFU	Analytical Sensitivity:	18 CFU/ cubic meter

Colony Identifications	Colony Forming Units Counted	Colony Forming Units/cubic meter
Alternaria alternata	2	36
Penicillium expansium	2	36
Other Fungi	5	91

The total colony forming units per cubic meter is 163.

Sample Number:	CMI060809-2-3	Incubation Temperature:	27 C
Sample ID:	Air Sample Foyer	Sample Volume:	55.19 L
Sample Condition:	Intact	Sample Type:	
Minimum Reporting Limit:	1 CFU	Analytical Sensitivity:	18 CFU/cubic meter

Colony Identifications	Colony Forming Units Counted	Colony Forming Units/cubic meter
Cladosporium cladosporioides	2	36
Alternaria alternata	1	18
Other Fungi	3	54

The total colony forming units per cubic meter is 108.

assuredbio

228 Midway Lane, Suite B
Oak Ridge, TN 37830
www.assuredbio.com
info@assuredbio.com
(865)813-1700

General Chain of Custody

AB Identifier: (for internal use only) CM1060809-0

Project Name: A Moldy House

Project Number: 123456

Project Date: 6/7/2009

	Inside	Outside
%RH	68	55
Temperature	72	85

Inspector: Certified Mold Inspector

Address: Best Mold Inspetors
123 Main Street
Anytown, TN 37770

Phone: 888-222-1111 **E-mail:** certified1@bmi.com

Turn Around Time	Sample Codes (SC)		Outside Conditions (Circle One)	
Direct Exam : 24 hours	SP- Spore Trap	BL- Bulk	C- Clear	TS- Thunderstorm
ERMI/DNA: 72 hours	SW- Swab	DU- Dust	R- Rain	S- Snow
Cultures: 7-14 days	TL- Tape Lift		W- Wind	

Sample ID	Description	SC	Total Volume/Area	Comments	MoldScan	ViaScan Genus Id Fungi	ViaScan Species Id Fungi	ViaScan Bacterial Colony Count	ViaScan Species Id Bacteria	ERMI	ARMI	Survey of Indoor Molds	Pathogenic Fungus Panel	Aspergillosis Assessment	Coliform Test	MPN	Legionella	Allergen Assay	Other:
	Basement Air Sample 1	BL	55.19 L		☐	☐	☑	☐	☐	☐	☐	☐	☐	☐	☐	☐	☐	☐	☐
	Air Sample Basement 2	BL	55.19 L		☐	☐	☑	☐	☐	☐	☐	☐	☐	☐	☐	☐	☐	☐	☐
	Air Sample Foyer	BL	55.19 L		☐	☐	☑	☐	☐	☐	☐	☐	☐	☐	☐	☐	☐	☐	☐
					☐	☐	☐	☐	☐	☐	☐	☐	☐	☐	☐	☐	☐	☐	☐
					☐	☐	☐	☐	☐	☐	☐	☐	☐	☐	☐	☐	☐	☐	☐
					☐	☐	☐	☐	☐	☐	☐	☐	☐	☐	☐	☐	☐	☐	☐
					☐	☐	☐	☐	☐	☐	☐	☐	☐	☐	☐	☐	☐	☐	☐
					☐	☐	☐	☐	☐	☐	☐	☐	☐	☐	☐	☐	☐	☐	☐
					☐	☐	☐	☐	☐	☐	☐	☐	☐	☐	☐	☐	☐	☐	☐
					☐	☐	☐	☐	☐	☐	☐	☐	☐	☐	☐	☐	☐	☐	☐
					☐	☐	☐	☐	☐	☐	☐	☐	☐	☐	☐	☐	☐	☐	☐
					☐	☐	☐	☐	☐	☐	☐	☐	☐	☐	☐	☐	☐	☐	☐

Relinquished By: Certified Mold Inspector Date: 6-7-09

Received By: [signature] Date: 6-8-09

9.8.1 The EPA's Mold Specific Quantitative Polymerase Chain Reaction (MSQPCR)

During the early 1990s a shift from standard polymerase chain reaction (PCR) analysis to quantitative PCR (QPCR) was well underway. Scientists working with the U.S. Environmental Protection Agency in Cincinnati, Ohio, recognized the usefulness of QPCR to quantify biological propagules, particularly mold spores. They quickly went to work and in the short span of seven years had developed over one hundred QPCR probes for the detection and quantification of mold species. They repackaged their technology and patented the probe and primer sequences that were developed (Vesper et al., 2005). The end product was a process and methodology to identify and quantify mold species, which they termed mold specific quantitative polymerase chain reaction or MSQPCR for short.

Once developed the EPA offered MSQPCR technology to environmental microbiology laboratories via a licensing agreement. However, not every lab was a suitable candidate to license the technology. The EPA stipulated that labs must meet both a science and business perspective. In other words labs had to have capable scientists who could carry out the complex MSQPCR analysis. Furthermore, labs had to submit a business plan to the EPA for review. The business plan had to detail how the laboratory was going to successfully commercialize the MSQPCR analysis. Needless to say only a handful of labs met the strict criteria set forth by the EPA. Most labs failed to meet the scientific specification that the EPA set. Today some 13 labs are licensed to conduct MSQPCR analysis, but only five labs offer the commercial assays (www.epa.gov/microbes/moldtech.htm).

9.8.2 The Environmental Relative Moldiness Index

The Environmental Relative Moldiness Index (ERMI) is an extremely robust application of MSQPCR technology for quantifying the mold burden of an indoor environment (Vesper et al., 2007a). Prior to the ERMI, laboratories would offer panels of MSQPCR probes to identify mold contamination indoors (Haugland et al., 2004). The problem with most MSQPCR panels, like spore traps, was that no scientific data was published to back the findings. This all changed with the ERMI.

The ERMI analysis was developed out of collaboration between the EPA and the U.S. Department of Housing and Urban Development (HUD) and funded under the Healthy Homes Initiative. The keystone ERMI study included some 1,100 homes from across the United States (Vesper et al., 2007a). Homes were randomly chosen using U.S. Census data. The ERMI represents a paradigm shift in indoor mold sampling. For one, the sample is a dust sample that is collected using any shop vac with a two-inch vacuum hose. A standard dust sampler is attached to the end of the shop vac hose and two $3' \times 6'$ areas are sampled for approximately five minutes to obtain one composite sample (http://assuredbio.com/learning-center/videos.html). The EPA's objective with the ERMI analysis was to design an analytical procedure in which it was extremely easy to collect samples while providing the best data available to determine mold contamination in a home.

The ERMI analysis has several things going for it that far exceed anything that has come before. The ERMI is an index that provides a score or value for the sampled home. The score is an indication of the mold burden in the home (Vesper et al., 2007a, 2009). Higher ERMI scores equate to higher mold burdens. Moreover, the EPA included language that specifies cut points, much like threshold values. The threshold values provide a means to provide homeowners with discrete delineations as to the level of mold contamination in the home (Figure 9.2). The ERMI was built using two groups of molds (Vesper et al., 2007a). Group I comprised 26 species of known water-intrusion molds; Group II included 10 species that are common outside but not indoors (Table 9.4). Again, the ERMI changes the paradigm in indoor mold testing. No longer is an outside sample needed. The outside sample is built into the ERMI analysis as the Group II species. To obtain an ERMI score, the data is log transformed and the Group II molds are subtracted from Group I molds. Doing so removes the effect of geographical region (log transformation) and removes the contribution of outside molds (subtraction of Group II molds). The ERMI score ends up being the true representation of the contribution or level of contamination of the Group I water-intrusion molds. These include species in the genera *Aspergillus*, *Penicillium*, *Chaetomium*, and *Stachybotrys* (Table 9.4).

The EPA only recommends the ERMI as an environmental test to determine the extent of mold contamination in a home; however, it has published several key papers that describe the relationship of the ERMI score to health (Vesper et al., 2006, 2007b, 2008). Of particular interest is the EPA's documentation of higher mold burdens correlating with early onset asthma in children (Vesper et al., 2007b). It turns out that the molds belonging to the Group I ERMI species are often allergenic and produce a variety of mycotoxins. It is not startling to find that these molds, when in high concentration, affect the respiratory health of home and building occupants.

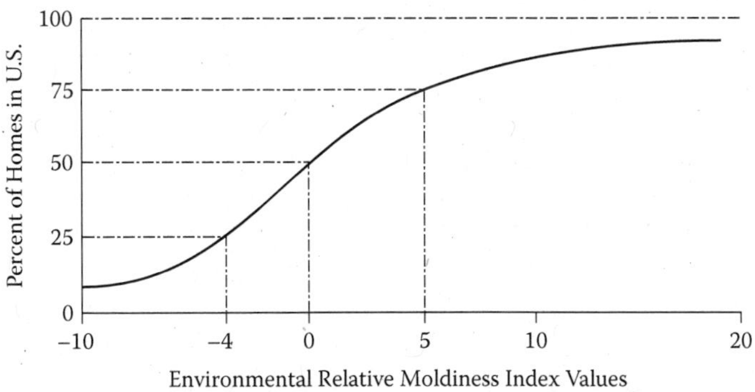

FIGURE 9.2 Mold burden threshold breaks as determined by the EPA. The bottom 25% represent the least moldy homes in the United States. The middle or 25%–75% categories represent the average mold burden in the United States. Above 75% are of the moldiest homes in the United States.

TABLE 9.4
Key to ERMI Assays

Assay Name	Target Species/Group of Species
Group 1 Molds	
Afumi	*Aspergillus fumigatus, Neosartorya fischeri*
Aochr1	*Aspergillus ochraceus/ostianus*
Arest	*Aspergillus restrictus/caesillus/conicus*
Asclr	*Aspergillus sclerotiorum*
Aungu	*Aspergillus unguis*
Avers2-2	*Aspergillus versicolor*
Apeni2	*Aspergillus penicillioides*
Cspha	*Cladosporium sphaerospermum*
Eamst	*Eurotium (Aspergillus) amstelodami/chevalieri/herbariorum/rubrum/repens*
Ppurp	*Penicillium purpurogenum*
Stac	*Stachybotrys chartarum*
Aflav	*Aspergillus flavus/oryzae*
Anigr	*Aspergillus niger/awamori/foetidus/phoenicis*
Asydo3	*Aspergillus sydowii*
Apull	*Aureobasidium pullulans*
Cglob	*Chaetomium globosum*
Pvari2	*Paecilomyces variotii*
Pbrev	*Penicillium brevicompactum/stoloniferum*
Pcory	*Penicillium corylophilum*
PenGrp2	*Penicillium crustosum/camembertii/commune/echinulatum/solitum*
Pspin2	*Penicillium glabrum/lividum/purpurescens/spinulosum/thomii*
Pvarb2	*Penicillium variable*
SCbrv	*Scopulariopsis brevicaulis/fusca*
SCchr	*Scopulariopsis chartarum*
Tviri	*Trichoderma viride/atroviride/koningii*
Wsebi	*Wallemia sebi*
Group 2 Molds	
Astrc	*Acremonium strictum*
Aaltr	*Alternaria alternata*
Cclad1	*Cladosporium cladosporioides* svar. 1
Cclad2	*Cladosporium cladosporioides* svar. 2
Cherb	*Cladosporium herbarum*
Austs2	*Aspergillus ustus*
Enigr	*Epicoccum nigrum*
Muc1	*Mucor amphibiorum/circinelloides/hiemalis/indicus/mucedo/racemosus/ramosissimus* and *Rhizopus azygosporus/homothalicus/microsporus/oligosporus/oryzae*
Pchry	*Penicillium chrysogenum*
Rstol	*Rhizopus stolonifer*

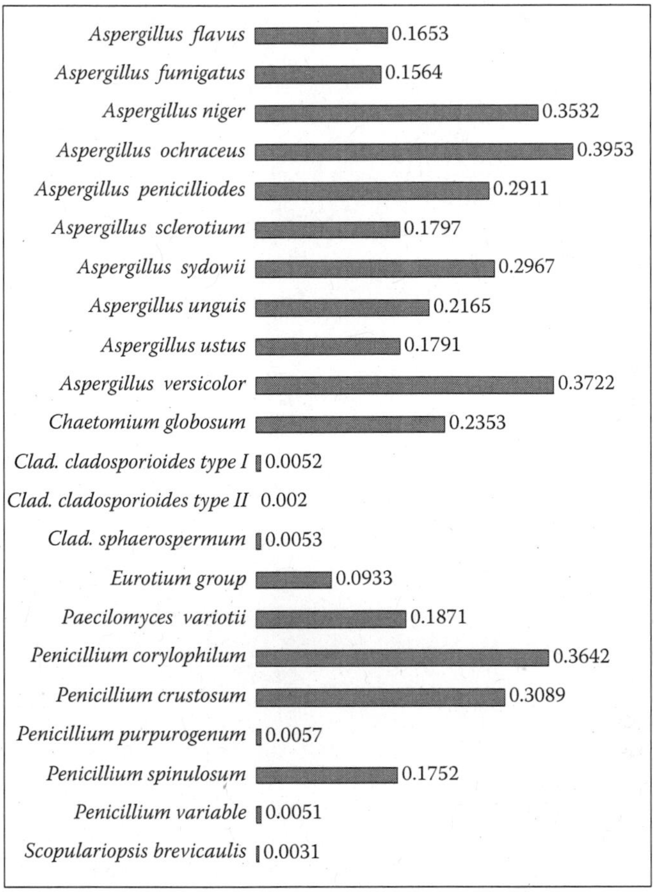

FIGURE 9.3 Species of mold detected by antibody Asp/Pen strip. Any value between 0.01–0.5 is a positive result.

9.9 ON-SITE MICROBIAL DETECTION TECHNOLOGIES

As advances in technology progress, instrumentation is becoming available for inspectors to make a presumptive determination of microbial contamination indoors. These tests are best used in a qualitative context to determine the presence or absence of a certain genera of concern or a critical biomass concentration that serves as an indicator of a possible indoor microbial situation.

9.9.1 ATP-Based Systems

Adenosine triphosphate, or ATP, is the currency that all living organisms use to run metabolism. ATP is the energy of life and hence is rapidly lost once an organism senesces. Technology that uses ATP to detect microbes relies on the fact that viable active cells are going to have an abundant supply of ATP, whereas dead cells will not.

A standard ATP detection system will consist of swabs, reagents, and a luminometer. Hygiena, a U.K. company, is widely known for its ATP detection devices (www.hygienausa.com). A swab is used for collecting samples. Once collected the swab is placed in a reagent mixture that contains a luciferase–luciferin complex. Luciferin is the chemical found in fireflies that allows them to emit light. When the luciferase–luciferin mixture is exposed to ATP luciferase, which is an enzyme, it activates luciferin and light is emitted. That light is detected in the luminometer. The luminometer has been calibrated using a ATP standard, hence it is possible to predict the concentration of ATP present in a sample. The concentration of ATP loosely correlates with the amount of microbial contamination.

The ATP method of analysis works well in the context of the pre- and postremediation process; where the ATP concentration is determined before cleanup and after cleanup. This concentration should be significantly lower after remediation is complete. However, one must be aware of the limitations of such tests. All organisms produce ATP; it is not just a microbial phenomenon. Blindly swabbing and measuring ATP concentrations in a home will likely mislead an inspector. The microbial source should be identified, then sampled and compared to postremediation sampling. ATP tells nothing of the types of organisms present. Laboratory analysis is recommended to identify whether bacteria or mold is the source of contamination and the genera or species present. Inspectors should be aware that dormant mold will contain very little ATP. ATP is related to metabolism and dormant mold maintains a metabolic stasis. Dormancy could cause false-negative results when a problem truly exists and needs to be addressed.

9.9.2 Antibody-Based Systems

Alexeter Technologies, LLC (www.alexeter.com) and Advnt Biotechnologies (www.advnt.org) offer antibody test strip systems for the detection of *Aspergillus/Penicillium* and *Stachybotrys* molds found in the EPA ERMI analysis (Table 9.4 and Figure 9.3). The Alexeter IAQ-Pro™ Asp/Pen test strip is a lateral-flow immunochromatographic device that uses two antibodies in combination to specifically detect the antigen in solution. One of the specific antibodies is labeled with a colloidal gold derivative. When sufficient antigen is present, the colloidal gold label provides a reddish-brown colored line that is visualized after accumulating in the test sample region on the device.

When a sample is added to the Alexeter IAQ-Pro Asp/Pen test strip, the sample begins to mix with the colloidal gold-labeled antibody and simultaneously moves along the strip membrane by capillary action. In the sample region of the test strip, if the target antigen is present, the secondary antibody captures the colloidal gold-labeled, primary antibody and bound antigen, forming a colored line or band in the sample (left side) window of the test strip. As an internal control, a second band visualized in the control (right side) window of the test strip is an indication that the test strip functioned properly. Two bands or colored lines are required for a positive result determination.

Alexeter IAQ-Pro Asp/Pen test strips display varying sensitivity levels to different mold types. Note that due to the unpredictable nature of environmental preparations, actual test sensitivity can vary for a given sample. In general, the IAQ-Pro Asp/Pen test is sensitive to mold species listed in the range of 1×10^5 mold spores per milliliter of sample. Since the test reacts against many different *Aspergillus/ Penicillium* mold spore types, a positive result may be an indication that the aggregate number of spores of different types exceeds the sensitivity level for the test. This test does not differentiate between the various mold types listed.

9.10 THE FUTURE OF INDOOR ENVIRONMENTAL SAMPLING AND TESTING

Mold testing has significantly morphed over the last decade. The advent of DNA diagnostics has significantly contributed to the recognition that indoor environmental microbiology is an important subdiscipline of microbiology that is worthy of scientific endeavor. Indoor environmental microbiology will continue to expand as a science over the next several decades. As more scientific studies address the adverse effects of mold and bacteria on human health, public awareness of indoor microbiology will increase. An aware and educated public will demand more sophisticated and rapid testing technology. Inspectors and laboratories will need to be prepared to serve the needs of their clients by providing rapid, onsite presumptive diagnostics coupled with confirmatory quantitative, species–specific laboratory diagnostics.

The movement toward green construction and the advancement of sensor technology is going to provide the opportunity to build preventive detection technology into new construction. Homes and buildings will have the hardware built in, much like a security system, that can notify offsite service providers, building owners, and insurance providers where and when a water-intrusion event has occurred. In advanced systems, if a pipe breaks, moisture sensors in walls will be able to notify the central control system to shut water off to piping in the immediate area. Likewise, sensors in roofs will be able to identify the exact location of a water leak. The cost savings to buildings, homeowners, and insurance companies from such technological advances will be enormous.

REFERENCES

Bloom, E., Grimsley, L. F., Pehrson, C., Lewis, J., and Larsson, L. 2009. Molds and mycotoxins in dust from water-damaged homes in New Orleans after Hurricane Katrina. *Indoor Air* 19: 153–158.

Deacon, J. W. 2006. *Fungal Biology*, 4th ed. Malden, MA: Blackwell.

Griffin, D. M. 1972. *Ecology of Soil Fungi.* London: Chapman and Hall.

Haugland, R. A., Varma, M., Wymer, L. J., and Vesper, S. J. 2004. Quantitative PCR analysis of selected Aspergillus, Penicillium and Paecilomyces Species. *Systematic and Applied Microbiology* 27: 198–210.

Heitman, J., Filler, S. G., and Edwards Jr., J. E., eds. 2006. *Molecular Principles of Fungal Pathogenesis.* Washington DC: ASM Press.

Macher, J., ed. 1999. *Bioaerosols: Assessment and Control.* Cincinnati OH: ACGIH.

Mueller, G., and Schmit, J. 2007. Fungal biodiversity: What do we know? What can we predict? *Biodiversity and Conservation* 16: 1–5.

Rhodes, J. C., and Brakhage, A. A. 2006. Molecular determinants of virulence in Aspergillus fumigatus, 333–345. In *Molecular Principles of Fungal Pathogenesis*, J. Heitman, S. G. Filler, J. E. Edwards, Jr. and A. P. Mitchell, eds. New York: Columbia University College Physicans and Surgeons.

Vesper, M. J., Vesper, S., Kahane, D., and Haugland, R. 2005. Mold-specific quantitative PCR: The emerging standard in mold analysis. *American Laboratory* : 10–12.

Vesper, S., McKinstry, C., Cox, D., and Dewalt, G. 2009. Correlation between ERMI values and other moisture and mold assessments of homes in the American Healthy Homes Survey. *Journal of Urban Health.*

Vesper, S., McKinstry, C., Haugland, R., Neas, L., Hudgens, E., Heidenfelder, B., and Gallagher, J. 2008. Higher Environmental Relative Moldiness Index (ERMIsm) values measured in Detroit homes of severely asthmatic children. *Science of the Total Environment* 394: 192–196.

Vesper, S., McKinstry, C., Haugland, R., Wymer, L., Bradham, K., Ashley, P., Cox, D., Dewalt, G., and Friedman, W. 2007a. Development of an environmental relative moldiness index for US homes. *Journal of Occupational and Environmental Medicine* 49: 829–833.

Vesper, S. J., McKinstry, C., Haugland, R. A., Iossifova, Y., Lemasters, G., Levin, L., Hershey, G. K. K., et al. 2007b. Relative moldiness index as predictor of childhood respiratory illness. *Journal of Exposure Science & Environmental Epidemiology* 17: 88–94.

Vesper, S. J. P., McKinstry, C. P., Yang, C. P., Haugland, R. A. P., Kercsmar, C. M. M. D., Yike, I. P., Schluchter, M. D. P., et al. 2006. Specific molds associated with asthma in water-damaged homes. *Journal of Occupational & Environmental Medicine* 48: 852–858.

Vesper, S. J. P., Varma, M. P., Wymer, L. J. M. S., Dearborn, D. G. M. D. P., Sobolewski, J. M. S., and Haugland, R. A. P. 2004. Quantitative polymerase chain reaction analysis of fungi in dust from homes of infants who developed idiopathic pulmonary hemorrhaging. *Journal of Occupational & Environmental Medicine* 46: 596–601.

Zorman, T., and Jersek, B. 2008. Assessment of bioaerosol concentrations in different indoor environments. *Indoor and Built Environment* 17: 155–163.

10 Research and Development, Directions in Construction Practice, and Summary Recommendations

Walter E. Goldstein, PhD, PE

CONTENTS

10.1 GENERAL RECOMMENDATIONS

Solutions in regard to preventing and treating mold infestation should preserve human health first; and second, assure best circumstances for the economic well-being of the housing industry, the consumers, the remediation industry, the insurance industry, and the nation. Mold appears in structures due to a variety of circumstances. It is discovered, testing is done, treated (including removing part of the structure), and the structure may be reassembled. The experience is costly, can be traumatic, can affect health, and legal expenses may be incurred.

Mold is a complex subject. However, there are many recommendations and suggestions to be made in this chapter. If one is facing a mold infestation, the first thing to do is to identify experts and reliable people, and report the matter. This can include remediation specialists and also your insurance company, assuming you have arranged for coverage. You will likely need to get the insurance company's authorization to have the remediation specialists and plumbers visit.

If this is a plumbing problem, then the water intrusion must be stopped immediately. If you cannot do this yourself, then call a plumbing service (again one you have used and know to be good and reliable, or one recommended by your insurance company). Do not delay or ignore the matter. Even if the water leakage is not overt, the presence of mold, if observed, is a serious matter that can negatively affect your health and the health of others, and cause extensive destruction to your property. Information in Chapter 2, for example, points out the potential epidemic dangers to health. The material on mold biology in Chapter 3 describes the entity that will appear and grow if an infestation results. Chapter 4 presents information on harmful products biosynthesized by mold. The case histories in Chapters 6, 7, and 8, for example, illustrate well the harm that can come to property. Chapter 9 presents information on metrics to analyze for the presence of mold. Chapter 5, the mathematical model and example cases, presents a format that is suggested to begin to pull information together to derive solutions to mold infestation.

The aforementioned necessary steps also require prior action before a mold problem occurs. The first step is having insurance coverage. This includes reading your policy, making sure you understand it, have all the relevant policy amendments, and the understand extent of your coverage. It is important to clarify the extent of your coverage and amounts ahead of time. If you have questions, get clarification from your agent. If your insurance coverage involves several policies, for example, if you reside in a condominium or townhouse development, make sure you understand which insurance policy is covering your dwelling and its components. Further, make sure your homeowners association covenants, conditions, and restrictions (CC&Rs) do not in some way interfere with your insurance coverage. This is a complicated matter requiring some investigation, and may often be confusing, unclear, and ambiguous. It is better to try to understand this before a problem occurs. You may need some help getting this clarified. That is why it is important to have a good insurance company and skilled agent.

A second step is already being aware of a plumbing service. A third step is being aware of a remediation specialist. Parties in this book can be contacted for recommendations in your local area. If you are in special housing, then you may need to reach a governmental agency such as the U.S. Department of Housing and Urban Development (HUD), U.S. Environmental Protection Agency (EPA), or Occupational Safety and Health Administration (OSHA) depending on your circumstances.

If you are in a situation where mold has affected your health, and you need to see a physician, then a recommendation from your primary physician is appropriate. Preferably, a physician experienced in this kind of problem is desired. There are lists of such people available on the Internet, or you may possibly obtain a referral from friends or family. There are many physical and behavioral manifestations of mold affecting health. One should study them expeditiously and visit your health care professional in a knowledgeable manner. Again, one has to be proactive and protect oneself and their loved ones and not just depend on others who may not have the same sense of urgency that you do.

If you are in a position where you are in the middle of a mold problem, and you cannot get satisfaction from those you deem responsible, and need an attorney, then the best course is to be referred to one from business associates, friends, or relatives. It is useful to use the Internet to search for those attorneys who have been successful in such cases and see if they can be helpful under the specific terms acceptable to you (financial and otherwise). You can also get recommendations from local officials.

These are all necessary steps depending on your situation. The contributors of this book present much information on mold infestation and remediation practice. They have also tried to make suggestions for prevention that are exhaustive with the hope that those facing this problem or working in this area will gain some good ideas and be stimulated to take actions from reading them. These suggestions are both long term and short term.

For example, there are solutions involving agriculture that are of a longer term nature since trees that produce products, such as materials in drywall, take a long time to grow and a longer time to change their structure. Therefore, one approach involves changing the materials of construction that are susceptible to mold.

10.2 AGRICULTURAL SOURCE SOLUTIONS

From what kind of tree is wallboard (a building material) derived? Does that tree have fungal infestations? Which trees might exude an allopathic chemical?

An allopathic chemical is produced by a plant to counter predators such as mold. Such chemicals may be isolated by wallboard manufacturers, produced in excess, for example, by fermentation of microorganisms that have been modified or selected to produce the allopathic chemical. The allopathic chemical can then be formulated into the wallboard as part of its manufacture to retard mold growth.

Retarding growth may be preferred unless one can identify an allopathic chemical that kills mold and does not affect human or animal life. For example, American chestnut trees are not affected by stump-rotting mold. Why is that? Does chestnut blight mold affect other vegetation? Would there be any clues here in regard to retarding fiberboard (wallboard) mold (Money, 2007)?

Examinations in this area in regard to toxicity require convincing data that no harm can be done using such allopathic chemicals. Such data can be obtained using biochemical and biological tests, including those involving genetics. Such tests may have to be devised. The tests are an important component of research to develop an improved wallboard product.

Materials in wallboard that are susceptible to fungal digestion are derived from trees. A long-term solution is to create tree hybrids that will resist the fungus. The materials produced may also be allopathic chemicals. However, other entities that provide defense may be involved. Such defenses can in principal be introduced through direct genetic modification by insertion of genetic materials into young tree tissue using vectors such as DNA or virus. The changes can also be accomplished through breeding. Genetic modification may involve a breeding component as well as horticulture. The tree containing the genetic modification (imparted through breeding) must be shown to grow economically. Since tree growth is long term and genetic modifications require testing of efficacy and safety, this solution is long term and not immediately of practical value. However, the concepts considered in this long-term solution should be evaluated for the ideas they suggest for short-term benefit. Research is often successful because one approach often brings ideas that result in alternatively successful approaches.

A subset of the previous solution is to genetically alter trees used for wallboard so they endogenously produce enzyme inhibitors to mold. This is direct, since molds release enzymes to convert to vegetative form and to propagate. Study of enzyme inhibitors has long been an important research topic in the health care area. Such research can be assessed to see how findings may apply to bioagriculture.

Chinese and Japanese chestnut are not affected by mold to the extent they are choked off and die, so molds in these cases may be different (Money, 2007). Are there strains of wallboard mold that are less virulent and what characterizes them? What is the resistance phenomenon? Maybe one can breed modified molds and release them. In this regard, in Chapter 4, it was noted that research at the U.S. Department of Agriculture (USDA) reports developing strains of the fungus *Myrothecium verrucaria* to infect a plant pest, the flowering ornamental kudzu (Boyette, 2009). Release of the fungus is intended to destroy the invading kudzu plant. USDA workers noted the necessity to reduce or stop the production of trichothecenes. As noted in this book, trichothecenes are a dangerous class of products also produced by mold that infects wallboard. Research in the area to block trichothecene production may result in findings that transfer to the medical area for enterprising companies since such events often result from research.

10.3 FERMENTATION SOLUTIONS RELATED TO BIOAGRICULTURE

Fermentation solutions related to bioagriculture may be short to midterm. For example, the aforementioned USDA workers reported using a proprietary distilled water method called "spore washing" to prevent the mold from producing the trichothecenes. Something in the wash water in the spore state affected secondary metabolite production. Perhaps this is the basis of a treatment for wallboard to stop production of toxins. They also found that by using liquid rather than solid fermentation culture, they silenced the trichothecene production or reduced the production significantly. The basis for the silencing is not known or at least not reported. However, one possibility is the oxygen or other gas partial pressure is affecting mold secondary metabolite production as this is common in fermentation. Perhaps these methods inherently involve an inhibitory step or steps that can prevent mold production of trichothecene in wallboard to at least alleviate part of the problem. Perhaps the mold growth, or its sporulation, or its generation into vegetative form can be prevented through related inhibitory steps. An important question is how do we know the molds will not select and become worse than what is presently apparent?

Producing molds less virulent or productive is also long term in nature. However, producing chemicals that may be inhibitors of mold formation or production of secondary metabolites (products) may be more short term and practical as a consumer or professional application, either during manufacturing of wallboard or as an application following installation as a preventative or treatment. Another patent, dealing with genetic modification of a plant to resist trichothecene, mentions expression of an enzyme that can degrade classes of compounds such as trichothecene (Hohn et al., 2006). The enzyme, trichothecene 3-O-acetyltransferase acetylates at a C3-OH position that apparently reduces the trichothecene toxicity. Use of such an enzyme (produced through fermentation) imbedded in wallboard, for example, may represent an approach to reduce the mycotoxicity of fungi. It is not inconceivable that interested parties could develop and use an enzyme-based spray to eliminate a toxin in wallboard.

10.4 SOLUTIONS INVOLVING BIOCHEMICAL, MICROBIAL, ENZYME, OR CHEMICAL INHIBITOR APPROACHES FOR USE IN MANUFACTURE OR BY CONSUMERS AS A SPRAY

This can be a practical solution, for example, spraying an allopathic chemical (or enzyme as noted) onto wallboard before a water leakage and mold problem occurs. Since spores spread everywhere, one must assume they exist in and around wallboard. A solution has to be more than preventing moisture, since water is bound to arise sometime due to leaks. Given that a leak occurs, a desirable solution would prevent mold growth until the leak is corrected.

The spray must reach all parts of the wallboard and affected wood areas, and not be toxic . One also must be assured that the spray will reach all areas that might be (are) affected. Otherwise, such a tactic may merely hide the problem, creating an unuseful and dangerous solution. However this approach can be a short-term solution worthy of research. The chemical sprayed must have staying power, though reapplication is certainly a possibility. Chlorine dioxide as well as chloramine-type applications may be an example here, besides the allopathic chemical and enzymatic approaches already mentioned.

As noted, one might directly address digesting or altering harmful mold products such as trichothenes through action of bacterial or enzymes that have shown activity against such substances. Bacterial combinations often have multidigestive properties that affect lipids, carbohydrates, and proteins in an enhanced manner. An example of this is a drain cleaner formulation that is available commercially (see Novozymes Biologicals, 2009). Often enzymes themselves may be found in combinations that attack substances that are normally recalcitrant to such attacks (Klibanov, 1989). The extension of enzymes as a commercial or consumer treatment in this field may have promise and is a relevant direction for research.

Experience in formulating enzymes for consumer use is widespread as they have been incorporated in detergents, for example, for many years. The safety of enzymes has been considered and would be relevant for formulating a wallboard treatment product. It would be essential, of course, to remove or inhibit carbohydrase activity so the product would attack the mold and its products, and avoid attacking wallboard material, which is dominantly cellulosic.

The activity and mode of action of such treatments (binding, microbial, or enzyme) could be studied and developed through direct use of the mathematical model presented in Chapter 5. Similarly, the experience and technology from other chapters in this book could aid in developing such products.

There are scientific questions to answer in regard to the aforementioned approaches. For example,

- What are conditions that promote mold propagation in wallboard (or other materials) or inhibit that growth? Perhaps, best conditions can be set to inhibit growth and prevent mold product formation. Perhaps, mold product formation can be stopped, thereby lessening the damaging power of the mold. There may be engineering approaches to assure dryness. Perhaps mold and its products can be selectively destroyed.
- Screen enzyme inhibitors to see which may inhibit enzymes released by fiberboard mold by exposing vegetative mold or mold cells with and without cell walls to such inhibitors in an experimental design. Find low-cost and safe inhibitors to be placed in wallboard.
- Similarly, screen enzymes and microbes to see which may have selective activity for molds or mold products. This can be approached classically through screening of collections or through mutation and screening. The matter can also be approached through genetic modification of organisms.
- Sulfur and copper-based compounds kill rust, a fungus of coffee. Triadimefon is a fungicide that may not bother other plants. Perhaps this is a clue. However, this fungicide can bother humans in many ways. It may be difficult to find a fungicide that does not also harm man and animals. Others are metalaxyl, fosetyl-aluminum, and phosphite. Research should be conducted to see if there is a pattern among these compounds that can provide hints to solutions from research.
- Perhaps an RNA virus may be helpful that is imbedded in the wallboard to penetrate and destroy the mold as it propagates. Can the virus mutate and in a zoonotic fashion leap to man? Just as man and animals, bacteria, and plants are susceptible to virus, so too are molds. A virus that specifically attacks Stachbotrys, for example, may be a useful solution if imbedded in wallboard or may be released as a spray.
- Just as there are fungal resistance genes in Robusta that make it more resistance to rust fungus than Arabica, maybe such genes also exist in trees used to make wallboard. However, does that transcend to protecting wallboard because of production of a biochemical (back to the previous allopathic chemical subject)?
- However, the rust mutates and overcomes resistance genes. So something is needed that is more general and overpowering, while still being safe. The wallboard fungus might overcome resistance. However, perhaps several allopathic compounds can be identified and obtained, and used in a systematic rotation in the wallboard manufacturing process as one of the contributors (Goldstein) and colleagues used to rotate dairy cultures so they would be less susceptible to bacteriophage.
- Spray penetration must be via carrier fluid into wallboard to kill mold, inhibit vegetative growth, and sporulation. The carrier fluid would suspend the active agent and facilitate its penetration to reach all parts of the wallboard where it would remain in a dry state until activated by moisture.

10.5 BIOLOGICAL SOLUTIONS

A biological solution would be to alter the DNA of the mold so sexual and asexual combination results in molds that do not propagate well, do not release toxins, and do not cause health issues. The modified molds would be released for recombination with other molds or to overwhelm other molds. Release them for combination without causing an environmental nuisance and hazard. This development would require extensive study of safety and environmental impact.

What is the vegetative (asexual) and sporulation process (sexual) intrinsically (e.g., see Chapter 3)? Look for spots in pathways to inhibit mold or prevent it from growing. Retardation of growth characteristics and confirmation of inhibitory agents can be studied, for example, using the mathematical modeling approach provided in Chapter 5.

10.6 ANALYTICAL SOLUTIONS TO DETECT MOLD IN HOUSING

Testing for the presence of mold may involve trying to determine what is present behind a wall. Physical sampling can be uninformative and perhaps prone to error. New tests that can monitor the presence of mold where it cannot be seen may be useful. The mold may give off a volatile that can be detected perhaps remotely, or a fingerprint of the mold may be detected remotely. Devices such as electronic noses have been examined to detect compounds in very small quantities (Suslick et al., 2002). Perhaps such compounds can be detected due to their diffusion even very slowly through building materials.

Moisture behind a wall is detected by using a probe with a defined sensitivity. Having water detection devices available for routine checking for leaks in a home by homeowners would also be a good preventative idea.

Such a device can conceivably be developed and applied to mold detection as well if the mold is present perhaps in the vegetative state where effects of mold growth (for example, by-products) are not yet observable. Services presently exist to check for mold in housing. Such devices would help make such testing much better for the consumer and professional services that are employed to detect mold. The devices can be used before purchasing a home and before and after remediation.

Analytical solutions (based on DNA and genetic analysis) are extensively covered in Chapters 3 and 9.

10.7 ARCHITECTURAL AND ENGINEERING MODIFICATIONS TO EXISTING AND NEW HOUSING

One of the problems in mold infestation is spotting it before it becomes harmful (in Chapter 9, application of "smart sensors" to detect water is suggested). This can also mean spotting a water leak before it causes mold infestation. One can envision constructing houses so that aesthetically attractive access panels or small doors are installed at select locations where a water leak is likely to occur (in the wall behind a dishwasher, a clothes washer, a refrigerator, a piping manifold, or behind a toilet, for example). One could then peer into the space periodically with a flashlight and see if water leaks are apparent. Installing this in new housing could be an option that may even result in some insurance credit. Installing such an access in existing housing with a standard design in multiple locations may also be similarly beneficial. Obviously, there is an added cost for this. Further, such an access takes away available space in a room. However, when one considers the impact of a costly mold remediation and excess removal of walls, and so forth, such an option may be worth considering.

Approaches that improve forced ventilation behind walls can be considered for new housing design. The intention in this case is to facilitate evaporation to (in principle) remove water before spores can be activated to vegetatively propagate.

Further, sealants and caulking that are a barrier to water leaks and mold from showers, bath tubs, and sinks, for example, do not last very long. Some research is called for into developing and applying new types of sealants that are more resistant to degradation and last longer. This can apply to existing and new housing. Perhaps the caulking can incorporate safe biochemicals and biological agents covered earlier in this chapter.

The rate of appearance of mold and factors that can hinder its spread as a consequence of moisture involving building modifications, and use of space-age sealants and detection devices, for example, could be evaluated using the model of Chapter 5.

Components of houses, valve manifolds, ice makers in refrigerators, dishwashers, clothes washers, and hot water heaters, for example, wear out and may become damaged due to corrosion. Sometimes, this wear can lead to water leaks. Sometimes, vibrations due to tremors or settling of a home can loosen pipe joints causing leaks. One has to be alert and watch for such problems. It is useful to understand the age of items in your home, have information on life expectancy, and perhaps estimate when to invest in replacement.

For example, plastic piping valve manifolds may have a 10-year life expectancy and may only be guaranteed for that period. The warrantee on the manifold may not go into effect until failure occurs. Of course, a failure leads to water intrusion and mold that may be undiscovered for an extensive period. Perhaps, in this example, it would be better to see about replacing the manifold proactively before its failure and seeing if the firm providing the warranty will still at least partially cover replacement.

It may be desirable to install soft water systems if the ion exchange treatment that produces softened water will minimize corrosion or metal and not accentuate it. Highly demineralized water can be corrosive due to oxidation effects. The water should be demineralized so its oxidation and reduction potential is minimized (Pourbaix, 1965). It may be that the water supplied to your home is perfectly adequate and softening is not required.

10.8 APPLICATION OF THE MATHEMATICAL MODEL (CHAPTER 5)

The mathematical model in Chapter 5 is designed to model mold growth, spore transformation, and product formation in a physical situation that mimics actual infestation. The model should be run and tested against specific situations. It can be used to decide how to optimally remediate, how to best prevent mold, and how to examine product formation and entities most injurious to health. Through development of applications, the model can be useful in selecting pharmaceuticals and treatment options to aid the physician.

REFERENCES

Boyette, C. 2009. Formidable fungus goes toe to toe with kudzu. *Agricultural Research*, 57(9): 4–5.

Hohn, T., Peters, C., and Saleron, J. May 23, 2006. Trichothecene Resistent Transgenic Plants. Assignee: Syngenta Participations, AG, U.S. Patent 7,049, 421.

Klibanov, A. 1989. "Enzymatic Catalysis in Anhydrous Organic Solvents." *Trends in Biochemical Sciences* 14(4) (April): 141–144.

Money, N. P. 2007. *The Triumph of the Fungi*, 15–16, 19–20. New York: Oxford University Press.

Novozymes Biologicals. 2009. Biochem 1000DL with BioS3112. Salem, VA: NovoZymes Biologicals Inc.

Pourbaix, M. 1965. A comparative review of electrochemical methods of assessing corrosion and the behavior in practice of corrodible material. *Corrosion Science* 5: 677–700.

Suslick, K., Rakow, N. A., and Sen, A. 2002. Colorimetric artificial nose having an array of dyes and method of artificial olfaction. U.S. Patent 6,495,102, December 5.

Index

C

Q

R